Andreas Wörner

A Metric Splitting of Alexandrov Spaces

Andreas Wörner

A Metric Splitting of Alexandrov Spaces

Boundary Strata of nonnegatively curved
Alexandrov Spaces and a Splitting Theorem

Südwestdeutscher Verlag für
Hochschulschriften

Imprint
Any brand names and product names mentioned in this book are subject to trademark, brand or patent protection and are trademarks or registered trademarks of their respective holders. The use of brand names, product names, common names, trade names, product descriptions etc. even without a particular marking in this work is in no way to be construed to mean that such names may be regarded as unrestricted in respect of trademark and brand protection legislation and could thus be used by anyone.

Publisher:
Südwestdeutscher Verlag für Hochschulschriften
is a trademark of
Dodo Books Indian Ocean Ltd., member of the OmniScriptum S.R.L Publishing group
str. A.Russo 15, of. 61, Chisinau-2068, Republic of Moldova Europe
Printed at: see last page
ISBN: 978-3-8381-1941-0

Zugl. / Approved by: Münster, WWU, Diss., 2010

Copyright © Andreas Wörner
Copyright © 2010 Dodo Books Indian Ocean Ltd., member of the OmniScriptum S.R.L Publishing group

Für Nelly

Wahrlich es ist nicht das Wissen, sondern das Lernen,
nicht das Besitzen sondern das Erwerben,
nicht das Da-Seyn, sondern das Hinkommen,
was den grössten Genuss gewährt.

Carl Friedrich Gauß
(Brief an Wolfgang Bolyai, 2. September 1808)

Summary

The main result of the present work is that a compact Alexandrov space of nonnegative curvature whose boundary consists of several strata such that the intersection of all strata is empty splits as a metric product.

More precisely, let M be an n-dimensional ($2 \leq n < \infty$) compact Alexandrov space with nonempty boundary ∂M. The boundary decomposes in a unique way into its primitive components, i.e. $\partial M = \bigcup_i P_i$ with $\dim P_i = n - 1$. We define boundary strata as unions of the components P_i. For a stratification of ∂M we always choose the boundary strata such that no set P_i is contained in distinct strata. Thus, a decomposition of ∂M into strata is in general possible in several ways.

We obtain and will prove the following splitting result.

Theorem. *Let M possess boundary strata $F_1, \ldots, F_{k+1}$ with $k \geq 1$ such that $F_1 \cap \ldots \cap F_{k+1} = \emptyset$, but all intersections of k strata are nonempty. (In other words, k is assumed minimal with the property $F_1 \cap \ldots \cap F_{k+1} = \emptyset$.) Then M is isometric to a product $S \times D$ of nonnegatively curved Alexandrov spaces, where $\dim S = n - k$, $\dim D = k$ and S is isometric to each intersection of k boundary strata.*

The main part is to prove this metric splitting under the assumption that there are no additional boundary strata, i.e. $\partial M = F_1 \cup \ldots \cup F_{k+1}$. In this case, ∂S turns out to be empty. We will formulate these statements as our Splitting Theorem 3.1 on page 43. Afterwards, this version of the Theorem extends easily to the general case formulated as Theorem 4.3 on page 68. There we will allow additional boundary strata. If there are some, they induce a stratification of ∂S.

The present work is set up like follows.

Summary

In the Introduction a short survey on known splitting theorems and other results in this context will be given; not only for Alexandrov spaces, but also for Riemannian manifolds with nonnegative curvature.

In Chapter 1 we collect basic facts about Alexandrov spaces and fix prerequisites and notation. This chapter contains no new results or proofs and can be skipped if Alexandrov geometry is known.

Chapter 2 provides tools required later. Not all results are new, but proofs are always included. The first sections begin with known results adapted to our needs. The subsequent sections contain results more and more specialized to the setup of our Splitting Theorem, but still may be of some independent interest.

In Chapter 3 the proof of the Splitting Theorem is carried out. We first give its statement and an outline on the proof. Afterwards, the proof is given in several steps, organized as theorems and propositions of their own. First, a fibration of M into subsets, called souls, is proved. The name is inspired by the fact that M can be retracted onto these souls. Thus, they turn out to be isometric. Finally, it is established that they are equidistant. Therefore, the product structure of M follows, and the S-factor is nothing but such a soul.

In the short Chapter 4 we give some corollaries and easy consequences of the Splitting Theorem. In particular, the following fact (which is probably well-known but nowhere written up) is (re)proved: A compact n-dimensional Alexandrov space has at most $2n$ boundary strata, and equality holds only for an Euclidean cuboid.

Acknowledgments

The work on this thesis was started with the intention to answer a question suggested by my supervisor Professor Burkhard Wilking. His conjecture as well as his ideas for a proof have been highly descriptive from the very beginning. For all years of development his guideline and confidence have been a great help and motivation to me. I am deeply grateful for his support and the opportunity to find the answer, albeit not as straightforward as expected. Frankly, I admit to have done so on purpose, since mathematical problems prevented (sometimes) exhausting discussions about soccer.

Furthermore, I would like to thank all members of the differential geometry group for fruitful conversations and a great atmosphere. In particular, I thank Professor Christoph Böhm for a more independent viewpoint. I also thank Professor Anton Petrunin for spending a lot of his time on explaining subtle facts in Alexandrov geometry to me (little and big ones).

Although not working in geometry, my friend of many years Dr. Heiko Dietrich was helpful as always and receives many thanks. Likewise, I thank my friends for their support. I am especially grateful for my parents encouraging me and always standing by.

Highest gratefulness, however, is dedicated to my personal muse and an angel inspiring and protecting me. Thank you for being there, Nelly.

Contents

Summary   iii

Acknowledgments   v

Introduction   1
   Metric splittings 1
   The Soul Theorem 3
   Extremal subsets and boundary strata 4

1 Basics about Alexandrov spaces   7
   1.1   Motivation 7
   1.2   Geometric properties 8
   1.3   Local properties 10
   1.4   Volume and dimension 13
   1.5   Differential and gradient 14
   1.6   Structure properties 17

2 Tools for the Splitting Theorem   23
   2.1   Shortest paths and zero measure 23
   2.2   Extendable shortest paths 25
   2.3   Boundary strata and doubling 27
   2.4   Superlevel sets 33
   2.5   Structure results on intersections of boundary strata 36

Contents

3 The Splitting Theorem   **43**
    3.1 Formulation 43
    3.2 Outline on the proof 44
    3.3 First case: two strata 45
    3.4 Fibration into souls 47
    3.5 Isometry of the souls via gradient flow 51
    3.6 The dual fibers 57
    3.7 Equidistance of the souls 61

4 Consequences of the Splitting Theorem   **67**

Bibliography   **71**

List of symbols   **75**

Index   **79**

Introduction

Metric splittings

Among splitting theorems for Riemannian manifolds with lower curvature bounds, there is the classical result by V.A. Toponogov in [Top 64]. It states that a complete Riemannian manifold M^n with nonnegative curvature which contains a straight line (i. e. a geodesic which is isometric to $\mathbb{R}$) is isometric to $M^{n-1} \times \mathbb{R}$, where M^{n-1} is again of nonnegative curvature. Here and throughout this work, we always mean sectional curvature; other kinds of curvature will be explicitly named. It turned out that the splitting property is induced by pure metric properties of M^n. In fact, if M^n is an Alexandrov space of nonnegative curvature, A.D. Milka proved in [Mil 67] that Toponogov's Splitting Theorem carries over. A proof in English can be found in [BBI 01, Section 10.5]. Loosely speaking, an Alexandrov space of nonnegative curvature is an inner metric space where Toponogov's Comparison Theorem is assumed to hold; we also assume completeness. More information about Alexandrov spaces will be given in Chapter 1, together with references.

In order to generalize Toponogov's Splitting Theorem, J. Cheeger and D. Gromoll proved in [CG 71] that in the case of Riemannian manifolds, nonnegative Ricci curvature is sufficient. For Alexandrov spaces, Y. Mashiko obtained in [Mas 02] a condition which is equivalent to the property that M^n splits into $M^{n-1} \times \mathbb{R}$, where M^n and M^{n-1} have the same lower curvature bound $\kappa \leq 0$. Namely, M^n splits as stated if and only if there is a nonconstant affine function $\varphi \colon M^n \to \mathbb{R}$ which is twice differentiable in a certain sense. This result in turn was generalized by S.B. Alexander and R.L. Bishop in [AB 05]. They proved—among other things—that if an Alexandrov space M^n with lower curvature bound κ admits a nonconstant K-affine function φ with $\kappa \leq K$, then M^n splits as a cone over an Alexandrov space. Here, a *cone* is some warped product of the space with an interval I. The warping function $g\colon I \to \mathbb{R}_+$ has to satisfy the differential equation $g'' + Kg = 0$. Moreover, φ restricted to any geodesic has to satisfy the same differential equation to be called K-affine. This is the condition if M^n has no boundary; otherwise the property has to hold on the doubling $\bar{M}^n$. We do not go into further detail here, but one sees that if $K = 0$ and g is constant, then Mashiko's result is recovered. For all other cases, Alexander and Bishop give detailed information as well.

In all results mentioned above, the spaces were always assumed to have finite dimension, indicated by the notation M^n. A finite dimensional Alexandrov space is locally compact (see Chapter 1) and, as a matter of fact, Milka's Splitting Theorem was formulated for locally compact Alexandrov spaces and needs no dimension assumptions. The question if the Theorem also holds for Alexandrov spaces which are not locally compact (and hence necessarily infinite dimensional), was answered affirmative by A. Mitsuishi in [Mit 10].

Still, all stated splitting theorems have in common that the space splits off some flat factor. The Theorem by Alexander and Bishop plays a special role, since warped products are involved. However, if we restrict to the cases of Euclidean products, also their result falls in the category of splitting theorems providing flat factors. Indeed, there are very few splitting theorems with non-flat factors—at least for spaces with lower curvature bounds. Gromoll and K. Tapp investigated in [GT 03] complete metrics of nonnegative curvature on a Riemannian manifold $M^2 \times \mathbb{R}^2$. A consequence of their results can be formulated according to Theorem 3.10 of the survey paper [Wil 07a] like follows: If $\mathbb{S}^2 \times \mathbb{R}^2$ carries a metric of nonnegative curvature which is not invariant under any effective action of a 2-torus, then it is a product metric (up to some diffeomorphism). Splitting results around the Soul Theorem by Cheeger and Gromoll are dealt with in the subsequent section.

For completeness, we mention here briefly that the situation is essentially different for spaces with upper curvature bounds. In fact, H.B. Lawson and S.T. Yau proved in [LY 72] for compact real analytic manifolds of nonpositive curvature that they split accordingly to their fundamental groups, provided the latter have no center. Independently, Gromoll and J.A. Wolf obtained in [GW 71] a similar result. Namely, if M is a manifold of nonpositive curvature and Γ is a group of homeomorphisms acting properly discontinuously by isometries on M and if $\Gamma\backslash M$ is compact, then a splitting of Γ induces some splitting of M. If, in addition, Γ has no center, the splitting carries over to the quotients. There are generalizations to Alexandrov spaces or, more precisely, to CAT(0) spaces. We also note that the results by Alexander and Bishop from above hold for both classes of Alexandrov spaces, those with lower and those with upper curvature bounds.

Introduction

The Soul Theorem

A source of theorems about nonnegatively curved manifolds splitting metrically is the Soul Theorem by Cheeger and Gromoll. They proved in [CG 72] the following fact: If M is a complete open manifold of nonnegative curvature, then there is a compact totally geodesic subset S—called soul—of lower dimension and without boundary, such that M is diffeomorphic to the normal bundle of S. The natural question arises, under which circumstances a metric splitting is induced. G. Walschap obtained in [Wal 88] that this is the case if S is simply connected and of codimension 2 and its normal bundle is flat. Then $M \cong S \times \mathbb{R}^2$, where $\mathbb{R}^2$ carries a metric of nonnegative curvature. Moreover, by independent results of M. Strake in [Str 88] and J.-W. Yim in [Yim 90], a metric splitting $M \cong S \times \mathbb{R}^{n-k}$ with $\dim S = k$ also occurs if the normal bundle of S has trivial holonomy group (without further assumptions on S).

In the proofs of these splitting results, the Sharafutdinov retraction plays an important role. In [Sha 79] V.A. Sharafutdinov constructed a 1-Lipschitz retraction map from M onto any soul S. In fact, S is not uniquely determined, and he also proved that given two souls $S, S' \subseteq M$, there is a diffeomorphism $M \to M$ mapping S isometrically onto S'. As part of the solution of the Soul Conjecture (i.e. S is a point and hence, M is diffeomorphic to $\mathbb{R}^n$, if M has at least one point of positive curvature), G. Perelman proved in [Per 94a] that the Sharafutdinov retraction coincides with the metric projection $\pi \colon M \to S$. Moreover, it is a Riemannian submersion of class C^1. The latter result was improved to C^2 by L. Guijarro in [Gui 00] and finally to C^∞ by B. Wilking in [Wil 07b]. In particular, π is a *submetry*, i.e. $\pi(\bar{B}_r(p)) = \bar{B}_r(\pi(p))$ $\forall p \in M$, $r \geq 0$. This property is again purely metric and one can ask, which facts around the Soul Theorem do carry over to Alexandrov spaces.

In his unpublished preprint [Per 91] Perelman constructed a Sharafutdinov retraction for Alexandrov spaces. Since the basic results are very important for the present work, we give some more detailed information here.

Let M be an n-dimensional Alexandrov space of nonnegative curvature with boundary. For the beginning, we assume that M is compact. Perelman showed that the distance function $d_{\partial M}$ is concave, i.e. its restriction to each shortest path is concave. This implies that the set $A \subseteq M$ where $d_{\partial M}$ attains its maximum is convex, hence an Alexandrov space. Clearly, $\dim A < n$. If $\partial A \neq \emptyset$, we iterate the procedure up to some convex subset without boundary. This subset then is a soul of M. A retraction map $M \to A$ was constructed directly by Perelman in [Per 91], but it is given more easily in terms of the later developed gradient flow. Loosely speaking, the gradient of some function $f \colon M \to \mathbb{R}$ at some $p \in M$ points in the direction where f increases most. This gives rise to gradient curves and a gradient flow; details will be given in Section 1.5. Thus, the gradient flow of $d_{\partial M}$ retracts M onto A, and so on. For non-compact Alexandrov

Introduction

spaces one may use combinations of Busemann functions instead of $d_{\partial M}$ in the first iteration step.

If M has positive curvature, $d_{\partial M}$ is strictly concave and hence, the soul is a point. However, it is not known yet, if positive curvature locally at at least one point is also sufficient. In other words, the Soul Conjecture is still open for Alexandrov spaces. Likewise, the Sharafutdinov retraction is not known to be a submetry.

The present work deals with cases where M is compact and more assumptions on the boundary ∂M are involved. Then a similar soul construction as above will give souls of a priori known dimension. Moreover, M will be fibrated into isometric souls, and (kind of) Sharafutdinov retraction maps will turn out to be submetries. This fact is crucial for the further result that the space in fact splits metrically into soul and another factor.

An exact formulation of our Splitting Theorem needs the term of *extremal subsets* introduced in the subsequent section.

Extremal subsets and boundary strata

A subset $E \subseteq M$ of an Alexandrov space with lower curvature bound is *extremal* if it is invariant under the gradient flow of any function d_p^2 with $p \in M$. The term was introduced by Perelman and A. Petrunin, see Section 1.6 for details. A single point $p \in M$ forms an extremal set if and only if there is no hinge xpy (i.e. xp and py are shortest paths) such that $\measuredangle xpy > \frac{\pi}{2}$. Equivalently, $\{p\}$ is extremal if and only if the space of directions Σ_p has diameter at most $\frac{\pi}{2}$. The boundary ∂M is an extremal set of locally constant codimension 1 and, vice versa, each extremal set of locally constant codimension 1 is part of ∂M. If an extremal set E contains no proper extremal set of the same dimension, then E is called *primitive*. Let M be compact; then the number of primitive extremal sets is finite. Hence, ∂M can be uniquely written as the finite union of all primitive extremal sets of codimension 1. For our purpose, however, the subdivision into primitive components is too restrictive.

We define a *boundary stratum* to be some union of primitive extremal subsets of codimension 1 and make our choices always in a way ensuring that two distinct boundary strata have no primitive component of full dimension in common. Let M be of nonnegative curvature (and still compact). A key point is now the observation that for any boundary stratum F the distance function d_F is concave. In particular, the case $F = \partial M$ from above is included. If we assume that ∂M consists of several strata, we may get more structure information about M. Indeed, the soul construction from above can be performed for each function d_F, where F is some boundary stratum. Moreover, this construction can be done not only inside M, but inside any superlevel set $d_{F'}^{-1}([s, a])$, where F' is another boundary stratum, $a = \max d_{F'}$ and $s < a$.

Introduction

Sharafutdinov retraction maps are obtained by the gradient flow of any function d_F. Thus, souls may be pushed around in M. In the setting of our Splitting Theorem (see below), this will imply that all souls are isometric and the Sharafutdinov retractions are submetries.

If M has positive curvature, the following results are due to Wilking in [Wil 06]. His Theorem 7 is formulated for orbits of Lie groups acting almost effectively on Riemannian manifolds. However, the orbit space is assumed to have positive curvature in the Alexandrov sense. The proof also uses Alexandrov techniques and essentially carries over. For these reasons, we may formulate the results here as follows.

A positively curved Alexandrov space M of dimension n has at most $n+1$ boundary strata. If equality holds, M is homeomorphic (also as a stratified space) to an n-simplex. If there are $k+1 < n+1$ boundary strata, M is homeomorphic to the join of a k-simplex and the intersection of all boundary strata.

We briefly sketch concepts of a proof here. Let $F_1, \ldots, F_{k+1}$ denote the boundary strata. Now, since M has positive curvature, all functions d_{F_i} are strictly concave. Therefore, the set $A_i := \{p \in M \mid d_{F_i}(p) \text{ is maximal}\}$ is just a point p_i and all gradient curves of d_{F_j}, $j \neq i$ end there. It follows that $p_i \in \bigcap_{j \neq i} F_j$. Consider p_{k+1}. By Perelman's Stability Theorem and averaging the function $d_{F_{k+1}}$ inside some small ball around p_{k+1}, one concludes that the tangent cone $T_{p_{k+1}}$ is homeomorphic to $M \setminus F_{k+1}$. Thus, $M \approx \bar{K}(\Sigma_{p_{k+1}})$, and since $p_{k+1} \in F_1 \cap \ldots \cap F_k$, the space $\Sigma_{p_{k+1}}$ satisfies the same assumptions as M. Now induction over the dimension proves the homeomorphism statements. The homeomorphisms respect the stratified structures, if the argument is refined using V. Kapovitch's relative version of the Stability Theorem.

The initial point for the present work was the question, if the homeomorphism statements carry over to nonnegative curvature. This clearly cannot be true, because if D is a (Euclidean) simplex with faces $F_1, \ldots, F_{k+1}$ and S is *any* nonnegatively curved Alexandrov space, then the product $M := S \times D$ has boundary strata $S \times F_i$, $i = 1, \ldots, k+1$ (and possibly other). In particular, the intersection of all strata $S \times F_i$ is empty, although $k < \dim M$.

This has led to the conjecture that such products are the only counterexamples.

Conjecture (solved by our Splitting Theorem). *Let M have nonnegative curvature and let F_1, $\ldots, F_{k+1}$ be boundary strata such that k is minimal with the property that $F_1 \cap \ldots \cap F_{k+1} = \emptyset$. Is it true that M is a product $S \times D$, where $\dim D = k$, $\dim S = n - k$ and S is isometric to each intersection of k boundary strata F_i?*

The affirmative answer to this question is the Splitting Theorem 3.1 on page 43. Chapter 3 is devoted to its proof, while Chapter 4 provides some further results as easy consequences. In particular, it turns out that the factor D has boundary strata $D \cap F_i$, and if M possesses further boundary strata apart from $F_1, \ldots, F_{k+1}$, they induce boundary strata of S. Otherwise,

S has no boundary. In Chapter 2 we provide tools being used in Chapter 3. Most of them hold in more general context and may be of independent interest. For instance, the fact that the intersection of k boundary strata has (locally constant) codimension k will be given there.

The easy case $k = 1$ (i. e. M has two boundary strata which do not intersect) can be proved more directly and without the machinery we will develop in Chapter 2. In fact, this case is so immediate that it is accepted as true, probably without being written up anywhere. In this case, the two boundary strata have to be primitive already. If they in turn have non-intersecting boundary strata, the Splitting Theorem can be used iteratively. We obtain that the number of boundary strata is bounded by $2n$, where $n = \dim M$. Equality holds if and only if M is a Euclidean cuboid. This result will be given in Theorem 4.5 on page 69.

In general, it is still a difficult open problem to give bounds for the number of (primitive) extremal sets and discuss rigidity. As mentioned above, the number is finite (recall that M is compact). It is conjectured that one can find an a priori upper bound only depending on $n = \dim M$. This problem is related to M. Gromov's estimate for the sum of Betti numbers of nonnegatively curved Riemannian manifolds, see [Gro 81]. In the latter case, the remaining conjecture is that a sharp bound is given by 2^n, attained by the torus. For Alexandrov spaces, the conjectured sharp bound for the number of primitive extremal sets is 3^n, attained by the cuboid. Besides the results given above for extremal sets of codimension 1, investigations have been made for 0-dimensional extremal sets, in other words, for extremal points. It can be shown in a rather elementary way, that there are at most 2^n such points. More complicated is a classification of all spaces attaining this maximal number. Recently, N. Lebedeva obtained solutions to this problem, but they have not yet been published. An overview will also be given in an upcoming book on Alexandrov geometry by Alexander, Kapovitch and Petrunin.

CHAPTER 1

Basics about Alexandrov spaces

1.1 Motivation

The classical Comparison Theorem by Toponogov states basically the following (see e.g. [GKM 68] or [CE 75]): Given a Riemannian manifold M^n with sectional curvature $\geq \kappa$ and a triangle $\triangle p_1 p_2 p_3$ in M^n whose sides $p_i p_j$ are shortest geodesics, then the corresponding triangle $\triangle \bar{p}_1 \bar{p}_2 \bar{p}_3$ with $|\bar{p}_i \bar{p}_j| = |p_i p_j|$ in the simply connected 2-dimensional space form of curvature κ is not "thicker". The latter can be expressed mathematically in different ways, for instance in terms of the corresponding angles (taking the form $\angle \bar{p}_i \bar{p}_j \bar{p}_k \leq \angle p_i p_j p_k$) or purely in terms of distances: For each $q \in p_i p_j$ and the corresponding $\bar{q} \in \bar{p}_i \bar{p}_j$ with $|\bar{q} \bar{p}_i| = |q p_i|$ the inequality $|\bar{q} \bar{p}_k| \leq |q p_k|$ holds. This enables us to take Toponogov's Theorem as a *definition* for spaces to have lower curvature bounds, and this definition is possible for the class of length spaces, not only for Riemannian manifolds. Furthermore, spaces with upper curvature bounds can be defined in the analogous way by reversing the inequalities in the angle and distance condition, respectively.

A comprehensive theory of spaces with lower curvature bounds and basic results were obtained by Y. Burago, Gromov and Perelman in [BGP 92]. Another approach (there exists some overlap, though) was given by C. Plaut in [Pla 91] based on concepts by A. Wald and V.N. Berestovskii. Plaut gives a detailed survey in [Pla 02], where later develops are also included, in particular the structure results by Perelman and Petrunin. An extensive introduction to the class of length spaces in general and spaces with curvature bounds in particular is given in the book [BBI 01] by D. Burago, Y. Burago and S. Ivanov. Since their basic setup as well as geometric tools and techniques trace back to A.D. Alexandrov, length spaces with curvature bounds are called Alexandrov spaces. In general the term may refer to both classes of spaces, those with lower and those with upper curvature bounds.

Throughout this work we consider always Alexandrov spaces with lower curvature bounds. In the next sections we collect basic results on such spaces without proofs. This survey is neither in chronological order nor in order of implication (and far from being complete), but it collects the results and tools needed later sorted by issues.

1.2 Geometric properties

First we introduce some notation in comparison geometry.

1.1 Definition. Let (X, d) be a length space[1]. For $n \in \mathbb{N}$, $\kappa \in \mathbb{R}$ we denote by S_κ^n the n-dimensional simply connected space form of curvature κ. Let $a, b, c \in X$ be pairwise distinct such that there are shortest paths ab, ac, bc.
For the triangle $\triangle abc$ we define the *comparison triangle* $\tilde{\triangle}_\kappa abc$ as the triangle $\triangle \bar{a}\bar{b}\bar{c}$ in S_κ^2 (provided it exists and is unique up to isometry) such that corresponding sides have equal lengths.
For the angle $\angle abc$ (again, provided it exists) we define the *comparison angle* $\tilde{\angle}_\kappa abc$ as the corresponding angle $\angle \bar{a}\bar{b}\bar{c}$ in the comparison triangle.

Now a curvature bound can be defined as follows.

1.2 Definition. A length space X has lower curvature bound κ if each point $p \in X$ has a neighborhood U such that for each triangle $\triangle abc$ in U the comparison triangle $\tilde{\triangle}_\kappa abc$ satisfies the following distance condition:
$|\bar{a}\bar{r}| \leq |ar|$ for each $r \in bc$ and the corresponding point $\bar{r} \in \bar{b}\bar{c}$ (i.e. $|\bar{b}\bar{r}| = |br|$).

As mentioned before, there are equivalent definitions, e.g. via angle comparison. Note that for general length spaces angles do not always exist (an angle can be defined in terms of lengths using the law of cosine and taking some limit), but in a space with curvature bound it turns out that angles always exist. All equivalent definitions and proofs can be found in [BGP 92] or in [BBI 01], for instance.

An immediate and important observation is that in spaces with lower curvature bounds shortest paths do not branch. Another implication is that angles are lower semi-continuous, i.e. for uniformly converging shortest paths $a_i b_i \to ab$ and $b_i c_i \to bc$ we have that $\angle abc \leq \liminf_{i \to \infty} \angle a_i b_i c_i$.

---
[1]a space with (strictly) intrinsic metric, i.e. between any two points exist paths whose lengths approximate (assume) the distance between the points.

1.2. Geometric properties

The comparison conditions are always local, so the question is, if, provided they hold locally, they also hold "in the large", i. e. for each triangle. This is true for complete spaces with lower curvature bounds and is known as the Globalization Theorem. (The analog theorem for upper curvature bounds needs more assumptions.) In the 2-dimensional case it was proved by A.D. Alexandrov, in the case of Riemannian manifolds by Toponogov and in the general case by Perelman (compare [BBI01, Section 10.3]). Mostly, it is known as the Toponogov Comparison Theorem, and also throughout this work we will refer to this name whenever some of the equivalent comparison conditions is used.

1.3 Theorem (Toponogov's Comparison Theorem). *If X is a complete length space with lower curvature bound κ then X has curvature $\geq \kappa$ also in the large.*

Here we will always consider complete length spaces with lower curvature bounds. In most cases we assume some more additional things.

1.4 Definition. The term *Alexandrov space* will always refer to a complete length space with lower curvature bound. For $n \in \mathbb{N}$ and $\kappa \in \mathbb{R}$ we denote by $\text{ALEX}^n(\kappa)$ the class of Alexandrov spaces with the following properties:

(i) The Hausdorff dimension is n, in particular finite;

(ii) the lower curvature bound is κ.

In addition, each space is

(iii) path-connected (except for $n = 0$);

(iv) strictly intrinsic (i. e. any two distinct points are connected by some shortest path whose length coincides with the distance of the points);

(v) locally compact;

(vi) second countable (equivalent to separable).

For $n = 1$, $\kappa > 0$ we assume in addition that the diameter does not exceed $\frac{\pi}{\sqrt{\kappa}}$.

Note that these properties are not at all independent, e. g. the local compactness is implied by the finite dimension. If, in turn, the Hausdorff dimension is assumed to be finite, it is an integer. In fact, the purely geometric definition of Alexandrov spaces implies many nice results on the local structure as we will see later.

For Alexandrov spaces of nonnegative curvature there is an extension of the distance comparison theorem due to U. Lang and V. Schröder. Here we only state a very special case of their theorem [LS 97, Theorem A] providing a powerful comparison tool.

1.5 Theorem. *For $n, m \in \mathbb{N}$ let $M \in \text{ALEX}^n(0)$ and let $f\colon S \to \mathbb{R}^m$ be a 1-Lipschitz map from some arbitrary subset $S \subseteq M$. Then there exists a 1-Lipschitz extension $\bar{f}\colon M \to \mathbb{R}^m$ of f.*

One sees immediately that this Theorem implies the distance comparison statement for spaces with nonnegative curvature.

As a last remark in this section we point out that the Gromov-Hausdorff limit of converging Alexandrov spaces with curvature $\geq \kappa$ is again an Alexandrov space with curvature $\geq \kappa$, since the completeness carries over as well as the comparison conditions.

1.3 Local properties

A Riemannian manifold M^n possesses at each point $p \in M^n$ a tangent space $T_p M^n$ as a local n-dimensional linear approximation to M^n. In particular, the set of unit vectors in $T_p M^n$ forms a space of directions such that two differentiable curves starting at p have the same direction if and only if they enclose an angle of value zero. Moreover, there is a geodesic starting at p in each arbitrarily given direction.

Since in an Alexandrov space M angles enclosed by shortest paths always exist, one can define that two shortest paths starting at $p \in M$ have the same direction if and only if they enclose a zero angle. This naturally defines a metric space, which is in general not complete. The completion is called the *space of directions* Σ_p at p. By taking the completion, directions not coming from shortest paths are created, so the statement from above about geodesics in Riemannian manifolds does not carry over to Alexandrov spaces. Nevertheless, if $\xi \in \Sigma_p$ is a direction not coming from any shortest path, there is a sequence $\xi_i \to \xi$ such that each ξ_i comes from some shortest path. This is sufficient for many applications.

However, many other results can be obtained only if there is at least some replacement for geodesics in all directions, for instance to define an exponential map. This can be achieved by gradient curves or by so-called quasigeodesics. The latter were introduced (in their generalized form) by Perelman and Petrunin in [PP 95] and studied with numerous applications by Petrunin, see e. g. [Pet 97]. A detailed proof of existence can be found in [Pet 07]. Quasigeodesics exist in finite-dimensional Alexandrov spaces for arbitrary initial data and have comparison properties almost like shortest paths.

Another approach to examine a metric space M locally at $p \in M$ is to take the so-called Gromov-Hausdorff tangent cone, i. e. to rescale the metric centered at p and to consider the pointed Gromov-Hausdorff limit, provided it exists. This procedure can be understood as zooming into the space with center p. If $M \in \text{ALEX}^n(\kappa)$, all rescaled spaces are Alexandrov spaces with curvature bounds tending to zero. Therefore, the pointed limit is an Alexandrov

1.3. Local properties

space with nonnegative curvature. On the other hand, we have the space of directions Σ_p and can take the metric cone $K(\Sigma_p)$ over it. Then these two cones coincide, which gives in turn that $\Sigma_p \in \text{ALEX}^{n-1}(1)$. More precisely, we have the following (see [BGP 92, §7] or [BBI 01, Section 10.9]) statement.

1.6 Theorem. *Let $M \in \text{ALEX}^n(\kappa)$. Then Σ_p is compact for all $p \in M$ and $\Sigma_p \in \text{ALEX}^{n-1}(1)$. If $n = 1$, then Σ_p consists of one or two points.*
The Gromov-Hausdorff tangent cone exists for all $p \in M$ and coincides with the metric cone $K(\Sigma_p)$ over Σ_p.

Note that it is essential that M has finite dimension, since otherwise Σ_p is in general not compact. We introduce some more notation here.

1.7 Definition. If pq is a shortest path, $\Uparrow_p^q$ denotes the direction in Σ_p of this path starting at p. By $\Uparrow_p^q$ or, more general, by $\Uparrow_p^A$ we denote the set of directions in Σ_p of all shortest paths from p to the point q or to some closed subset $A \subseteq M$, respectively.
The tangent cone $K(\Sigma_p)$ is denoted by T_pM and, by abuse of standard definition, its elements are called *(tangent) vectors*. The apex of T_pM is denoted by o and length and scalar product of vectors are defined as follows.

$$|v| := |ov|$$
$$\langle v, w \rangle := \frac{1}{2}(|v|^2 + |w|^2 - |vw|^2) = |v| \cdot |w| \cdot \cos \angle vow$$

for $v, w \in T_pM$ (recall that T_pM has nonnegative curvature, so $\angle vow = \tilde{\angle}_0 vow$).

Now it is possible to define a logarithm map.

1.8 Definition. Let $M \in \text{ALEX}^n(0)$ and $p \in M$. For each $q \in M \setminus \{p\}$ fix some shortest path pq and let $\xi_q = \Uparrow_p^q$. Then the map

$$\log_p : M \to T_pM, \, q \mapsto \begin{cases} |pq| \cdot \xi_q & : q \neq p \\ o & : q = p \end{cases}$$

is called a *logarithm map* at p.

The Comparison Theorem implies that $\log_p$ is noncontracting, i.e. lengths are not shortened. For other curvature bounds $\kappa \neq 0$ replace $T_pM = K(\Sigma_p)$ by the κ-cone K_p^κ (compare [BBI 01, Example 10.2.2]) over Σ_p.

The fact that Σ_p is again an Alexandrov space, but of lower dimension, enables in many cases proofs via induction on dimension. As an example, for $M \in \text{ALEX}^n(\kappa)$ and $p \in M$ there exists a noncontracting map $f\colon M \to S^n_\kappa$ which maps shortest paths starting at p isometrically, see [BBI 01, Proposition 10.6.10]. Now assume that this is already proved up to dimension $n-1$ and hence for Σ_p. The noncontracting map $\varphi\colon \Sigma_p \to \mathbb{S}^{n-1}$ extends to the κ-cones, i. e. to a map $\tilde\varphi\colon K^\kappa_p \to S^n_\kappa$, and then $\tilde\varphi \circ \log_p$ is the desired map.

On the other hand, also definitions are possible in an inductive way. An important example is the notion of the boundary of an Alexandrov space.

1.9 Definition. Let $M \in \text{ALEX}^n(\kappa)$. The *boundary* ∂M of M is defined as follows. For $n=1$ let ∂M be the topological boundary (note that M is a segment, a line or a circle). For $n > 1$ some point $p \in M$ is a boundary point $p \in \partial M$ if Σ_p has nonempty boundary.

More information about the boundary is given in Section 1.6.

As we have seen, the tangent cone $T_p M$ is a generalization of the tangent space. We do not have $T_p M = \mathbb{R}^n$, but at least $\mathbb{R}^n$ is a comparison space for $T_p M$ in the sense of curvature bounds. Nevertheless, points where the tangent cone is in fact isometric to Euclidean space are of special interest.

1.10 Definition. Let $M \in \text{ALEX}^n(\kappa)$. A point $p \in M$ with tangent cone isometric to $\mathbb{R}^n$ (or equivalently $\Sigma_p \cong \mathbb{S}^{n-1}$) is called a *regular point*. Non-regular points are called *singular*.

It turns out that the singular points form in fact a very spare set. Y. Otsu and T. Shioya proved in [OS 94, Theorem A] that it has Hausdorff dimension $\leq n-1$, in particular, it is of n-dimensional measure zero. Thus, the set of regular points is dense. Moreover, it is a countable intersection of open sets and, due to the result [Pet 98, 1.10] of Petrunin, it is a convex set.

A weaker condition than being regular is being so-called (n, δ)-strained for $\delta > 0$. Loosely speaking, an (ℓ, δ)-strained point possesses 2ℓ directions which are nearly perpendicular or opposite, respectively. In the precise definition, not directions but shortest paths are used and comparison angles instead of angles in order to overcome some problems with incontinuities. The following definition is according to [BBI 01, Definition 10.8.9].

1.11 Definition. For $M \in \text{ALEX}^n(\kappa)$ a point $p \in M$ is called (ℓ, δ)-*strained* if M possesses ℓ pairs of points (a_i, b_i) such that

$$\tilde\angle a_i p b_i > \pi - \delta \quad \text{and} \quad \tilde\angle a_i p a_j,\ \tilde\angle a_i p b_j,\ \tilde\angle b_i p b_j > \frac{\pi}{2} - 10\delta$$

for all $i \neq j$. The collection $\{(a_i, b_i)\}$ is called an (ℓ, δ)-*strainer* for p.

It is convenient to take all δ "small enough", e. g. $\delta \leq \frac{1}{100n}$. Then each (n, δ)-strained point has a neighborhood U which is bi-Lipschitz homeomorphic to an open region in $\mathbb{R}^n$ (see

[BGP 92, Theorem 5.4] or [BBI 01, Theorem 10.8.18]). In particular, this holds for all regular points, because a point is regular if and only if it is (n, δ)-strained for all $\delta > 0$. Moreover, at regular points the bi-Lipschitz constant can be assumed arbitrarily close to 1 by choosing the neighborhood U small enough.

1.4 Volume and dimension

The following theorem is a consequence of the last statement of the previous section, see [BGP 92, §6] or [BBI 01, Section 10.8].

1.12 Theorem. *A finite-dimensional Alexandrov space is locally compact.*
The Hausdorff dimension $\dim_H M$ of a locally compact Alexandrov space M is an integer or infinity. Moreover, $\dim_H$ is locally constant, i.e. each open neighborhood of M also has Hausdorff dimension $\dim_H M$.

The last part of the Theorem follows from the fact that in an Alexandrov space M for $0 < \lambda < 1$ one can define a λ-homothety centered at some point p. If M has nonnegative curvature, this mapping is noncontracting. More precisely, for $p \in M$ and λ given, fix for each $x \in M$, $x \neq p$ some shortest path px and define $f(x) \in px$ such that $|pf(x)| = \lambda|px|$. Then it follows immediately from Toponogov's Comparison Theorem that f is noncontracting. Therefore, two metric balls centered at p have equal Hausdorff dimension, namely $\dim_H M$. The Bishop-Gromov inequality (compare [BGP 92, Theorem 10.2]) is proved in the same way.

If the curvature bound of M is $\kappa < 0$, one can restrict the λ-homothety to some ball $B_R(p)$ and obtains a co-Lipschitz map with a constant $c = c(\kappa, \lambda, R)$. Then the statement about the local dimension follows analogously as before. To prove the Bishop-Gromov inequality, however, another proof is necessary, see e.g. [BBI 01, Section 10.6].

Another application of strained points is to prove that the Gromov-Hausdorff limit A of Alexandrov spaces A_i with $\dim_H(A_i) = n$ $\forall i$ fulfills $\dim_H(A) \leq n$. All these results together lead to the Compactness Theorem by Gromov, see [BBI 01, Section 10.7].

1.13 Theorem (Gromov's Compactness Theorem). *For $n \in \mathbb{N}$ and $\kappa \in \mathbb{R}$ let $\mathfrak{M}(n, \kappa)$ denote the class of Alexandrov spaces M with curvature bound κ and $\dim_H(M) \leq n$. Then $\mathfrak{M}(n, \kappa)$ is boundedly compact, i.e. all closed bounded subsets are compact, with respect to Gromov-Hausdorff distance.*
In particular, for each $D > 0$ the subclass $\mathfrak{M}(n, \kappa, D) := \{M \in \mathfrak{M}(n, \kappa) \mid \operatorname{diam} M \leq D\}$ is compact.

As we have seen in this section, property (v) of Definition 1.4 on page 9 (i.e. local compactness) follows in fact from property (i) (i.e. finiteness of dimension). According to [BBI 01,

Proposition 2.5.22], a locally compact length space is boundedly compact. In particular, all closed balls are compact and therefore the space can be exhausted by a countable sequence of compact sets. Thus, the space is separable or, equivalently, second countable (property (iv) of Definition 1.4).

1.5 Differential and gradient

Let $M \in \text{Alex}^n(\kappa)$ and $f\colon U \to \mathbb{R}$ be a function defined on an open subset $U \overset{\circ}{\subseteq} M$. We want to have a notion of differentiability for f and a differential $d_p f$ defined on the tangent cone $T_p M$ with $p \in U$. In order to do this, f can be restricted to some shortest path $pq \subseteq U$ and differentiated in the usual way. If $\gamma\colon [a, b] \to U$ is a shortest path parametrized by the arc length with $\gamma(a) = p$, $\gamma(b) = q$, then the directional derivative $d_p f(\uparrow_p^q)$ is given by $(f \circ \gamma)'(a)$, provided this (one-sided) derivative exists. This procedure is done for all directions in Σ_p coming from shortest paths, and if the obtained function extends continuously to the entire Σ_p, we say that f is *differentiable at p*. In order to obtain the differential $d_p f$ as a function on $T_p M$, we extend the directional derivatives positively homogeneous to $T_p M$. This means, if $v \in T_p M$ fulfills $v = \lambda \xi$ for some $\lambda > 0$, $\xi \in \Sigma_p$, then let $d_p f(v) := \lambda d_p f(\xi)$.

An important class of differentiable functions with explicit differential is given by distance functions (compare [BGP 92, §11]).

1.14 Lemma. *Let $A \subseteq M$ be a closed subset and $f\colon M \to \mathbb{R}$, $x \mapsto |xA|$. Then the differential $d_p f$ exists for all $p \in M \setminus A$ and the following holds:*

$$d_p f(\xi) = -\cos|\Uparrow_p^A \xi| \quad \forall \xi \in \Sigma_p$$

Another possibility to obtain the differential is via blow-ups. Recall that $T_p M$ can be considered as the pointed limit of rescaled spaces centered at p. Evaluating f on these blown up spaces near p and taking the differential quotients (if existing) also gives the differential $d_p f$, see e.g. [Pet 07, Definition 1.3.1]. This procedure can be generalized to maps between metric spaces with certain local structures, in particular to maps between Alexandrov spaces. These generalizations together with a more general and powerful definition of tangent cones via ultra limits were extensively studied by A. Lytchak in [Lyt 05] and [Lyt 06]. Among other results, Lytchak also gives a general Rademacher Theorem for Lipschitz maps between spaces with curvature bounds, see [Lyt 05, Theorem 1.6].

Moreover, in the case of Alexandrov spaces the differential does not only exist almost everywhere, but is linear. To make this precise, we give the following version of Rademacher's Theorem:

1.5. Differential and gradient

1.15 Theorem. *Let $X \in \text{ALEX}^n(0)$, $Y \in \text{ALEX}^k(0)$ and let $f \colon X \to Y$ be a Lipschitz map. Then for almost all $p \in X$ the following holds.*

- *$T_p X$ is isometric to $\mathbb{R}^n$.*
- *$T_{f(p)} Y$ is isometric to $\mathbb{R}^m \times C$, where $C \in \text{ALEX}^{k-m}(0)$ is a cone.*
- *$d_p f$ exists and is a linear map with image in the $\mathbb{R}^m$-factor of $T_{f(p)} Y$.*

For a proof see [Lyt 02, Proposition 3.8] or [Lyt 05, Corollary 1.5]. Indeed, since almost all points $p \in X$ are regular and differentiability is a local property, one can assume that $X = \mathbb{R}^n$.

Furthermore, the Theorem still holds if f is locally Lipschitz on neighborhoods of almost all points.

Curves are maps[2] of special interest. We introduce some notation here.

1.16 Definition. For an Alexandrov space M let $\gamma \colon [a, b] \to M$ be a curve. The *right tangent vector* $\gamma^+(t)$ with $t \in [a, b)$ and the *left tangent vector* $\gamma^-(t)$ with $t \in (a, b]$ are defined as follows.
$$\gamma^\pm(t) := \lim_{\varepsilon \to 0+} \left(\frac{1}{\varepsilon} \log_{\gamma(t)} \gamma(t \pm \varepsilon) \right)$$
if the limits exist.

It is clear by definition that if γ is a shortest path, the right and left tangent vector exist and are opposite (i.e. $|\gamma^+(t)| = |\gamma^-(t)|$ and $\langle \gamma^+(t), \gamma^-(t) \rangle = -1$) for all $t \in (a, b)$. If γ is a Lipschitz curve, the same is true for almost all t (according to [PP 95, §2] or by Rademacher's Theorem, whose proof is similar) and the length of γ is given by
$$L(\gamma) = \int_a^b |\gamma^+(t)| \, dt.$$

Now we want to introduce the gradient of a function and gradient curves. In Riemannian geometry (as well as in ordinary analysis) the direction of the gradient is at each point uniquely determined as the direction in which the function increases most. Conversely, this property can be taken as a definition of the gradient. In order to do this, we have to restrict to a class of functions ensuring both, the existence of the differential and the uniqueness of the direction where the function increases most.

A function $\varphi \colon [a, b] \to \mathbb{R}$ is called λ-*concave* if the function $t \mapsto \varphi(t) - \frac{\lambda}{2} t^2$ is concave on $[a, b]$. Let $M \in \text{ALEX}^n(\kappa)$ without boundary and $U \overset{\circ}{\subseteq} M$ an open subset. A locally Lipschitz

---

[2] In this work, we consider curves as maps, i.e. we fix a parametrization. Hence, the terms "curve" and "path" are interchangeable.

function $f\colon U \to \mathbb{R}$ is called λ-concave, if the restriction to each shortest path contained in U is λ-concave. For spaces M with boundary consider all shortest paths in the doubling $\bar{M}$ (i. e. two copies of M are glued together along their boundaries), take their canonical projection onto M and restrict f to these curves. Details on the doubling procedure are given in Section 2.3.

1.17 Example. For $M \in \text{ALEX}^n(0)$ and $p \in M$ the function $x \mapsto \frac{1}{2}|xp|^2$ is 1-concave on M. This is a direct consequence of the Comparison Theorem (in fact, it enables an equivalent *definition* of the curvature bound) and the fact that the doubling $\bar{M}$ lies again in $\text{ALEX}^n(0)$.

1.18 Definition. Let $M \in \text{ALEX}^n(\kappa)$ and $U \overset{\circ}{\subseteq} M$. A function $f\colon U \to \mathbb{R}$ is called *semiconcave* if for each point $p \in U$ there is some $\lambda(p) \in \mathbb{R}$ and an open neighborhood of p in which f is $\lambda(p)$-concave.

It is clear that a semiconcave function $f\colon U \to \mathbb{R}$ is differentiable at each point $p \in U$ and the differential $d_p f$ is a concave function on $T_p M$. Thus, the gradient which is defined as follows exists and is well-defined.

1.19 Definition. Let $M \in \text{ALEX}^n(\kappa)$ and $f\colon U \to \mathbb{R}$ be semiconcave with $U \overset{\circ}{\subseteq} M$. For $p \in U$ the vector $\nabla_p f \in T_p M$ fulfilling the following conditions (compare [Pet 07, Definition 1.3.2])

(i) $d_p f(v) \leq \langle \nabla_p f, v \rangle \quad \forall v \in T_p M$;

(ii) $d_p f(\nabla_p f) = |\nabla_p f|^2$

is called the *gradient* of f at p.

It follows that $\nabla_p f$ in fact points in the direction where f increases most, provided it increases in first order in some direction. If not, $\nabla_p f = o \in T_p M$. In [PP 95, §3] that was used as the definition for the direction of $\nabla_p f$ and the following facts were proved, compare also [Pet 07, 1.3.3–1.3.5].

1.20 Lemma. For $M \in \text{ALEX}^n(\kappa)$, $U \overset{\circ}{\subseteq} M$ and a λ-concave function $f\colon U \to \mathbb{R}$ let $p, q \in U$ such that $pq \subseteq U$. Then the following holds.

(i) $\langle \uparrow_p^q, \nabla_p f \rangle + \langle \uparrow_q^p, \nabla_q f \rangle \geq -\lambda |pq|$.

(ii) $|\nabla_p f|$ *is lower semi-continuous, i. e.* $\liminf\limits_{p_i \to p} |\nabla_{p_i} f| \geq |\nabla_p f|$.

These results enable the construction of gradient curves. For their definition we follow [Pet 07, Definition 2.1.1], while in [PP 95, 3.2] a reparametrized version is used. The latter has the property that the parameter of the gradient curve coincides at each point with the value of the function at that point. The first version, however, is probably more standard and gives more direct geometric results.

1.21 Definition. For $M \in \text{ALEX}^n(\kappa)$ let $f\colon M \to \mathbb{R}$ be a semiconcave function. A curve $\alpha\colon I \to \mathbb{R}$ is called *gradient curve* of f if

$$\alpha^+(t) = \nabla_{\alpha(t)} f \quad \forall\, t \in I.$$

For each point $p \in M$ there is a unique f-gradient curve starting at p and existing for infinitely long future time, while the past is in general not determined. The construction is also possible for infinite dimensional Alexandrov spaces. A proof is given in [PP 95, Appendix]. Provided some function $f\colon X \to \mathbb{R}$ has a gradient at each point and the lower semi-continuity (ii) of Lemma 1.20 holds, Lytchak proved that gradient curves for f exist even if X is a locally compact metric space, see [Lyt 06, Proposition 1.6]. In this general setting the gradient curves are not necessarily unique. Uniqueness is obtained under the assumption of more local properties, see [Lyt 06, Proposition 9.1].

The existence and uniqueness of gradient curves gives in turn the notion of some gradient flow, sometimes also called gradient push. The latter term stresses out the fact that the past is not determined.

1.22 Definition. Let $M \in \text{ALEX}^n(\kappa)$ and $f\colon M \to \mathbb{R}$ be a semiconcave function. Let α_p denote the f-gradient curve starting at p for each $p \in M$. For $t \geq 0$ the map

$$\Phi_f^t\colon M \to M,\; p \mapsto \alpha_p(t)$$

is called the *gradient flow* of f.

Among many nice properties of the gradient flow, see e.g. [Pet 07, Section 2.2], we only point out a particular one which is important for later use.

1.23 Lemma. *Let $M \in \text{ALEX}^n(\kappa)$ and $f\colon M \to \mathbb{R}$ be a λ-concave function with $\lambda \in \mathbb{R}$. Then Φ_f^t is $e^{\lambda t}$-Lipschitz. In particular, if $\lambda = 0$, then f is nonexpanding (i.e. 1-Lipschitz) for all $t \geq 0$.*

This follows from Lemma 1.20(i). Indeed, if gradient curves α_p, α_q start at $p, q \in M$ and we set $d(t) := |\alpha_p(t)\,\alpha_q(t)|$, the formula mentioned implies that $d'(t) = -\langle \uparrow_p^q, \nabla_p f \rangle - \langle \uparrow_q^p, \nabla_q f \rangle \leq \lambda \cdot d(t)$ and hence $d(t) \leq e^{\lambda t}$.

1.6 Structure properties

As we have seen in Section 1.3, Alexandrov spaces have local geometric properties similar to those of Riemannian manifolds, if both have finite dimension, say n. A major difference is,

Chapter 1. Basics about Alexandrov spaces

however, that Riemannian manifolds look in this sense the same at each point. The tangent space is everywhere isomorphic to $\mathbb{R}^n$ and each point has a neighborhood homeomorphic to $\mathbb{R}^n$. The tangent cone of an Alexandrov space $M \in \text{ALEX}^n(\kappa)$ is almost everywhere isometric to $\mathbb{R}^n$, and near such points M is an n-dimensional topological manifold. Simple examples show that in general not the entire space M is a manifold, but there are topological singularities. The question arises, if the set of these singularities is again almost everywhere a topological manifold, of course one of dimension $\leq n - 1$, and so on.

An affirmative answer was given by Perelman in the unpublished preprint [Per 91, Theorem 0.1] and more elaborated in [Per 94b]. By developing and using Morse theoretic arguments Perelman proved that each point $p \in M$ has a neighborhood U which is homeomorphic to the tangent cone $T_p M$. In addition, U can be chosen as a spherical neighborhood, i.e. as a metric ball $B_r(p)$. This result enables a stratification of M whose strata are topological manifolds. In this form it does not take geometric singularities into account. In order to obtain a stratification also adapted to the geometric structure, Perelman and Petrunin introduced in [PP 94] so-called extremal subsets of M. Although they are not Alexandrov spaces themselves, they inherit the topologically stratified structure of M via some relative Morse Lemma. Thus, the stratification of M can be refined such that geometric singularities are respected.

In the following we give a short overview on the basic results in this structure theory of Alexandrov spaces. Apart from the cited references above, we mention Kapovitch's paper [Kap 07] on Perelman's Stability Theorem, the highlight of all those structure results.

1.24 Definition. For $M \in \text{ALEX}^n(\kappa)$ a subset $E \subseteq M$ is called *extremal* if for any semi-concave function $f \colon M \to \mathbb{R}$ the gradient flow does not leave E once it has reached it, i.e. $\Phi_f^t(p) \in E \quad \forall p \in E, t \geq 0$.

This definition is due to Petrunin in [Pet 07, Definition 4.1.1]. In Theorem 4.1.2 he proved that the definition is equivalent to the one given in [PP 94, Definition 1.1], which says that a closed subset $E \subseteq M$ is extremal if and only if for each point $q \in M \setminus E$ the following holds: If the restricted distance function $d_q\big|_E$ has a local minimum at $p \in E$, then $\nabla_p d_q = o$, i.e. the point p is critical for the distance function d_q on M.

There are numerous results on extremal subsets in the papers mentioned above. We will only stress out some which are essential for the present work.

1.25 Theorem. *Let $M \in \text{ALEX}^n(\kappa)$ and $E, F \subseteq M$ be extremal. Then the following holds.*

(i) $\Sigma_p E \subseteq \Sigma_p$ *is extremal for all $p \in E$.*

1.6. Structure properties

(ii) $E \cap F$, $E \cup F$, $\overline{E \setminus F}$ are extremal with the following spaces of direction.

$$\begin{aligned}
\Sigma_p(E \cap F) &= \Sigma_p E \cap \Sigma_p F, \\
\Sigma_p(E \cup F) &= \Sigma_p E \cup \Sigma_p F, \\
\Sigma_p(\overline{E \setminus F}) &= \overline{\Sigma_p E \setminus \Sigma_p F} \qquad \text{for all } p \in E \cap F,\, E \cup F,\, \overline{E \setminus F},\, \textit{respectively.}
\end{aligned}$$

Here $\Sigma_p E$ denotes the set of all directions in Σ_p coming from differentiable curves which lie in E. The Theorem expresses implicitly that $\Sigma_p E$ is well-defined, but of course this fact had to be proved. In addition, also the reverse of (i) holds: If a subset $E \subseteq M$ contains at least two points and $\Sigma_p E \subseteq \Sigma_p$ is extremal for all $p \in E$, then E is extremal. A single point p forms an extremal set if and only if $\operatorname{diam} \Sigma_p \leq \frac{\pi}{2}$.

1.26 Definition. Let $M \in \mathrm{ALEX}^n(\kappa)$. An extremal set $E \subseteq M$ is called *primitive* if it contains no proper extremal subset with nonempty relative interior. The *main part* $\mathring{E}$ of a primitive extremal set E is the subset of all points not lying in another primitive extremal set.

The next theorem gives the announced stratification of an Alexandrov space, compare [PP 94, 3.8].

1.27 Theorem. *Let $M \in \mathrm{ALEX}^n(\kappa)$ be compact. The number of extremal subsets in M is finite. Each extremal subset (in particular M itself) can be uniquely represented as the disjoint union of main parts of primitive extremal sets.*
Moreover, all these main parts are topological manifolds and hence, M is a topologically stratified space.

In the non-compact case, all statements hold inside any compact subset of M; in particular, the stratification property carries over. We will give some more detail on the Morse theoretic arguments developed by Perelman and used in order to prove the last statement of the Theorem above.

Like the tangent space of a Riemannian manifold was generalized by the tangent cone of an Alexandrov space, there is a class of topological spaces generalizing topological manifolds in a similar way. While the latter have at each point a neighborhood homeomorphic to Euclidean space (of fixed dimension), a so-called *MCS-space* (meaning *multiple conic singularities*) of dimension n is a topological space X such that each point $x \in X$ has a neighborhood pointed-homeomorphic to the open cone over its boundary (pointed means mapping x to the apex of the cone). This boundary in turn is a compact MCS-space of dimension $n-1$, and the empty set is defined to be the unique MCS-space of dimension -1.

In [Per 94b, Theorem III] Perelman proved that a finite-dimensional Alexandrov space is an MCS-space. In order to do this, we introduce a special subclass of semiconcave functions, see [Kap 07, Section 5].

1.28 Definition. For $M \in \text{ALEX}^n(\kappa)$ let $A_\alpha \subseteq M$ be a collection of closed subsets, $\lambda_\alpha \geq 0$ with $\sum_\alpha \lambda_\alpha \leq 1$ and $\varphi_\alpha \colon \mathbb{R} \to \mathbb{R}$ twice differentiable with $0 \leq \varphi'_\alpha \leq 1$. Then the function

$$f \colon M \to \mathbb{R}, \, p \mapsto \sum_\alpha \lambda_\alpha \varphi_\alpha\big(|pA_\alpha|\big)$$

is called *admissible*; more precisely, admissible on $U = M \setminus (\cup_\alpha A_\alpha)$.

It follows from the definition that an admissible function f is 1-Lipschitz and semiconcave on U and for the differential at $p \in U$ the following holds (by the chain rule and Lemma 1.14 on page 14):

$$d_p f(\xi) = \sum_\alpha -\lambda_\alpha \varphi'_\alpha\big(|pA_\alpha|\big) \cos |\!\uparrow_p^{A_\alpha} \xi| \quad \forall \xi \in \Sigma_p$$

The coefficients $a_\alpha := \lambda_\alpha \varphi'_\alpha\big(|pA_\alpha|\big)$ satisfy $a_\alpha \geq 0$, $\sum_\alpha a_\alpha \leq 1$. Let another function $g \colon M \to \mathbb{R}$ be admissible at p with a differential satisfying $d_p g(\xi) = \sum_\beta -b_\beta \cos |\!\uparrow_p^{B_\beta} \xi|$ for $\xi \in \Sigma_p$. Then some scalar product is defined by

$$\langle d_p f, d_p g \rangle := \sum_{\alpha, \beta} a_\alpha b_\beta \cos |\!\uparrow_p^{A_\alpha} \uparrow_p^{B_\beta}|.$$

Note that this scalar product is not completely well-defined as long as equal functions may have different representations according to Definition 1.28. For this reason we identify admissible functions by their representation.

A map $f \colon M \to \mathbb{R}^k$ is called admissible on $U \subseteq M$, if all its component functions f_i are admissible on U.

1.29 Definition. Let $f \colon M \to \mathbb{R}^k$ be admissible on $U \subseteq M$ and $\varepsilon > 0$. A point $p \in U$ is called *regular point for f* if there is some $\varepsilon > 0$ such that the following conditions hold.

(i) $\langle d_p f_i, d_p f_j \rangle < -\varepsilon \quad \forall i \neq j$;

(ii) there exists some $\xi \in \Sigma_p$ with $d_p f_i(\xi) > \varepsilon \quad \forall i$.

For an admissible map $f \colon U \to \mathbb{R}^k$ we also allow the subsequent composition with some bi-Lipschitz homeomorphism G between open sets in $\mathbb{R}^k$. The map $g = G \circ f$ is also called admissible and its regular points are the regular points for f.

1.6. Structure properties

The set of regular points is open, even with a fixed value for ε (this follows from [Kap 07, Lemma 5.2]). There are no regular points if $k > n = \dim M$. An admissible map is open near its regular points.

We can now formulate Perelman's Local Fibration Theorem or Morse Lemma, see [Per 94b, Theorem 1.4]; compare also [Kap 07, Theorem 6.8], [Pet 07, Section 8].

1.30 Theorem (Local Fibration Theorem). *Let $M \in \text{ALEX}^n(\kappa)$ and $g \colon M \to \mathbb{R}^k$ be admissible on $U \subseteq M$. If $p \in U$ is a regular point for g, then there exists an open neighborhood U_p of p such that $g|_{U_p}$ is a topological bundle map. More precisely, there is a homeomorphism $h \colon U_p \to (g^{-1}(g(p)) \cap U_p) \times g(U_p)$ satisfying $\pi_2 \circ h = g|_{U_p}$ (where π_2 is the projection onto the second factor) and the fiber $g^{-1}(g(p)) \cap U_p = \left(g|_{U_p}\right)^{-1}(g(p))$ is an MCS-space of dimension $n - k$.*
Moreover, if there is an open subset $V \subseteq U$ such that all points in V are regular and $g|_V$ is a proper map, then $g|_V$ is the projection of a locally trivial fiber bundle.

Since an admissible map restricted to the set of its regular points is an open map, the second statement of the Theorem follows from the first one by a result of L.C. Siebenmann in [Sie 72, Theorem 5.4, Corollaries 6.14, 6.9]. The first statement is proved via reverse induction on k starting with a trivial statement for $k = n + 1$, since in this case there are no regular points. The proof is carried out in [Per 94b], compare also [Kap 07, Section 6].

The Fibration Theorem applied to the constant map $M \to \mathbb{R}^0$ gives that M is an MCS-space. In addition, let M_k be the set of points in M possessing a neighborhood homeomorphic to $\mathbb{R}^k \times C$, where C is a cone and k has the maximal possible value. Then each set M_k is a topological manifold of dimension k, and the disjoint union yields the stratification of M.

1.31 Corollary. *The closure of each stratum M_k is an extremal subset. In particular, the boundary is an extremal subset, hence closed, and satisfies $\partial M = \overline{M}_{n-1}$.*

This follows from Theorem 1.30 and the equivalent definition for extremal subsets (given after Definition 1.24 on page 18), compare [PP 94, 1.2]. The statement about the boundary follows from [Per 91, Theorem 4.6] and the fact that ∂M has everywhere local dimension $n - 1$. This in turn follows by induction, since the codimension of ∂M in M locally at $p \in \partial M$ is 1 if and only if $\Sigma_p(\partial M) = \partial(\Sigma_p)$ has codimension 1 in Σ_p.

Theorem 1.30 can be extended to a relative version for extremal subsets. In order to do this, the restriction of admissible maps to extremal subsets is considered. Perelman and Petrunin proved in [PP 94] that the main results carry over, however with the following generalization of MCS-spaces: An $\widetilde{\text{MCS}}$-space of dimension $\leq n$ is defined analogously to an MCS-space, only that the cone at each point is taken over the compact boundary being an $\widetilde{\mathrm{MCS}}$-space of dimension $\leq n-1$. If in Theorem 1.30 the term "MCS-space of dimension $n-k$" is replaced by "$\widetilde{\mathrm{MCS}}$-space of dimension $\leq n-k$", the statement also holds for the restriction of g to some extremal subset. In particular, extremal subsets are $\widetilde{\mathrm{MCS}}$-spaces and hence, primitive extremal subsets are MCS-spaces, since their dimension is locally constant. This gives the stratification of M respecting the geometric structure.

Kapovitch gives more details on the Relative Fibration Theorem in [Kap 07, Section 9] in order to prove his relative version of Perelman's Stability Theorem. The latter is somehow a generalization of the Fibration Theorem in the sense that not only subspaces of a fixed Alexandrov space M are compared, but the (compact) Alexandrov spaces themselves. Perelman proved the Stability Theorem in [Per 91, Theorem 0.3].

1.32 Theorem (Stability Theorem)**.** *Let $M \in \mathrm{ALEX}^n(\kappa)$ be compact. Then there exists some $\varepsilon(M) > 0$ such that each compact Alexandrov space $N \in \mathrm{ALEX}^n(\kappa)$ with Gromov-Hausdorff distance $d_{GH}(M,N) < \varepsilon(M)$ is homeomorphic to M.*

The relative version due to Kapovitch, see [Kap 07, Theorem 9.2], also takes extremal subsets into account.

CHAPTER 2

Tools for the Splitting Theorem

2.1 Shortest paths and zero measure

Assume we have an Alexandrov space $M \in \text{ALEX}^n(0)$ and a subset $X \subseteq M$ which has n-dimensional Hausdorff measure zero. Given two points $p, q \in M$ we want to find a shortest path γ with endpoints close to p and q, respectively, such that $\gamma \cap X$ has 1-dimensional measure zero. In Euclidean space the existence of such shortest path is ensured by Fubini's Theorem. For Riemannian manifolds with lower Ricci curvature bounds the result follows from the segment inequality proved by J. Cheeger and T. Colding in [CC 96, Theorem 2.11].

In the case of Alexandrov spaces Otsu and Shioya considered in [OS 94] the set of singular points, which has measure zero (compare Section 1.3) and satisfies the condition from above, see [OS 94, Theorem 6.4]. In fact, they proved even more, since the set of singular points has special properties. In particular, since by Petrunin's result the set of regular points is convex, we can find our desired shortest path such that it does not intersect at all with the set of singular points.

Nevertheless, for an arbitrary subset X with measure zero we still can adopt the first part of the proof by Otsu and Shioya.

2.1 Proposition. *Let $M \in \text{ALEX}^n(0)$ and $X \subseteq M$ with $\mu_n(X) = 0$, where μ_n denotes the n-dimensional Hausdorff measure. Let $p, q \in M$ and $\varepsilon > 0$.*
Then there exists a shortest path $\gamma = \hat{p}\hat{q}$ with endpoints $\hat{p} \in B_\varepsilon(p)$, $\hat{q} \in B_\varepsilon(q)$ such that $\mu_1(\gamma \cap X) = 0$. Moreover, γ can be chosen to consist only of regular points.

Since the proof is according to [OS 94], all citations in the following proof refer to that paper, if not otherwise stated.

Proof. Denote the set of singular points in M by S_M. We have $\mu_n(S_M) = 0$ by Theorem A.

Let $x \in B_\varepsilon(p)$, $y \in B_\varepsilon(q)$ be regular points. Then there is a local bi-Lipschitz natural chart (Lemma 3.6(4)), i.e. there are points $p_1, \ldots, p_n \in M$ and an open neighborhood U of y such that the mapping $\varphi \colon U \to \mathbb{R}^n$, $y \mapsto (|yp_1|, \ldots, |yp_n|)$ is a bi-Lipschitz homeomorphism. Moreover, the point x can be taken as one of the base points p_i. In addition, we choose U small enough to ensure $U \subseteq B_\varepsilon(q)$.

For $t \in \mathbb{R}$ we set $F_t := \{u \in U \mid |xu| = |xy| + t\}$. The local bi-Lipschitz chart gives us the existence of some $\delta > 0$ such that for all t with $|t| < \delta$ we have that $0 < \mu_{n-1}(F_t) < \infty$ (since this is true for the hyperplanes $\varphi(F_t) \subseteq \mathbb{R}^n$). The coarea formula in $\mathbb{R}^n$ implies that

$$\int_{-\delta}^{\delta} \mu_{n-1}(F_t)dt \leq c_1 \cdot \mu_n(\{u \in U \mid \big||xu| - |xy|\big| < \delta\})$$

with a constant $c_1 > 0$ coming from the bi-Lipschitz chart φ. Together with $\mu_n(X \cup S_M) = 0$ we obtain that

$$\int_{-\delta}^{\delta} \mu_{n-1}(F_t \cap (X \cup S_M))dt \leq c_1 \cdot \mu_n\big(\{u \in U \mid \big||xu| - |xy|\big| < \delta\} \cap (X \cup S_M)\big) = 0.$$

Thus, we can choose some $t \in (-\delta, \delta)$ such that $\mu_{n-1}(F \cap (X \cup S_M)) = 0$ holds for $F := F_t$. We equip F with the induced metric from the ambient space M, rescale it by the factor $1/|xF|$ and consider the cone $K(F/|xF|)$. Let $A := \bar{B}_{|xF|}(o) \subseteq K(F/|xF|)$ be the closed ball of radius $|xF|$ centered at the apex. Now we define a map $f \colon A \to M$ like follows.

For each point $z \in F$ we fix a shortest path xz. Given $(s, a) \in A$ (i.e. $0 \leq s \leq |xF|$, $a \in F$) let $f(s, a)$ be the point $b \in xa$ with $|xb| = s$. For $a \in F$ let $\beta_a := [0, |xF|] \times \{a\}$ denote the ray segment in A. It immediately follows that f maps each β_a isometrically onto the corresponding shortest path xa. Moreover, by Toponogov's Comparison Theorem, the map f is noncontracting. Indeed, let $(s, a), (t, b) \in A$. Triangle comparison gives (using Theorem 1.5 on page 10) that $d(f(s, a), f(t, b)) \geq d(\overline{f(s, a)}, \overline{f(t, b)})$.

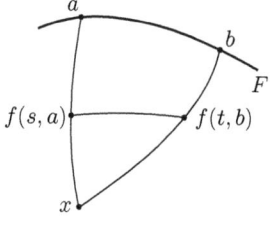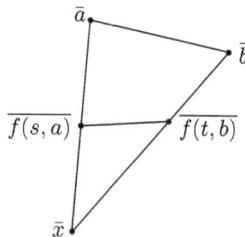

2.2. Extendable shortest paths

In addition, the inequality $d(\overline{f(s,a)}, \overline{f(t,b)}) \geq d((s,a),(t,b))$ follows from the definition of the cone metric on $K(F/|xF|)$. Hence f is noncontracting.

By the coarea formula (applied to the corresponding cone over the hyperplane $\varphi(F) \subseteq \mathbb{R}^n$) and $\mu_n(X \cup S_M) = 0$, we obtain for any $r \in (0, |xF|)$ that

$$\int_F \mu_1\big((\beta_z \cap f^{-1}(X \cup S_M)) \setminus B_r(o)\big) dz \leq c_2 \cdot \mu_n\big((f^{-1}(X \cup S_M)) \setminus B_r(o)\big)$$
$$\leq c_2 \cdot \mu_n\big((f(A) \cap (X \cup S_M)) \setminus B_r(x)\big) = 0$$

for some constant $c_2 > 0$ depending on the bi-Lipschitz chart φ and on Lipschitz constants coming from the central projections onto F or $\varphi(F)$, respectively, in the corresponding cones. Thus, for any $r \in (0, |xF|)$ and almost all $z \in F$ we have that

$$\mu_1\big((xz \cap (X \cup S_M)) \setminus B_r(x)\big) = 0.$$

A sequence $r_i \searrow 0$ gives now the result

$$\mu_1\big(xz \cap (X \cup S_M)\big) = \mu_1\left(\bigcup_i ((xz \cap (X \cup S_M)) \setminus B_{r_i}(x))\right) = 0$$

for almost all $z \in F \subseteq B_\varepsilon(q)$.

In addition, because of $\mu_{n-1}(F \cap (X \cup S_M)) = 0$, the point $z \in F$ can be chosen regular. Then, by the convexity of the set of regular points, the shortest path xz consists entirely of regular points. □

2.2 Extendable shortest paths

As stated earlier, Otsu and Shioya even proved that $\dim_H(S_M) \leq n - 1$. In order to give a simpler proof for the weaker statement $\mu_n(S_M) = 0$, Otsu introduced in [Ots 97] the cut locus Cut_p for $p \in M$ and proved (Proposition 2.2 op. cit.) that $\mu_n(\text{Cut}_p) = 0$. More precisely, Cut_p is defined as the set of all points $x \in M$ such that there is no shortest path starting at p and containing x as an interior point. Thus, for all $y \notin \text{Cut}_p$ a shortest path py can be extended beyond y, in particular it is unique.

Given a closed subset $A \subseteq M$ we are interested in the set of points $x \in M \setminus A$ such that a shortest path between x and A can be extended beyond x. We ask if this is true for almost all $x \in M \setminus A$, which is affirmative by Otsu's result in the case $|A| = 1$ and hence also if A is countable. A modification of Otsu's proof gives a positive answer for arbitrary closed subsets.

2.2 Proposition. *Let $M \in \text{ALEX}^n(0)$ and $A \subseteq M$ be a closed subset. Let X denote the set of points $x \in M \setminus A$ such that each shortest path between x and A cannot be extended as a shortest path beyond x. Then X has n-dimensional Hausdorff measure zero.*

Proof. For each $\delta > 0$ we define the following set:

$$W^\delta := \{\, x \in M \mid x \text{ lies on a shortest path from some } \tilde{x} \text{ to } A \text{ with } |\tilde{x}A| = (1+\delta)|xA| \,\}$$

Since the corresponding point $\tilde{x}$ is unique for $x \in W^\delta$, the map

$$E^\delta : W^\delta \to M\,,\ x \mapsto \tilde{x}$$

is well-defined. In the case $|A| = 1$ it is an immediate consequence of Toponogov's Comparison Theorem that E^δ is a $(1+\delta)$-Lipschitz map. In our general case we want to apply the extended Comparison Theorem 1.5 on page 10 and have to ensure that the triangle in $\mathbb{R}^2$ we want compare with exists. For that reason we only prove the existence of the Lipschitz constant locally and for $\delta \leq \frac{1}{4}$. This is no restriction since $W^{\delta'} \supseteq W^\delta$ holds for $\delta' \leq \delta$.

Given $z \in M \setminus A$ we set $U := B_r(z) \cap W^\delta$ with $r = \frac{1}{4}|zA|$. For $x, y \in U$ let p, q denote the endpoints of the shortest paths from $\tilde{x} = E^\delta(x)$ and $\tilde{y} = E^\delta(y)$, respectively, to A. We obtain that

$$\begin{aligned}
|\tilde{x}z| + |xz| &\leq |\tilde{x}x| + 2|xz| = \delta \cdot |xA| + 2|xz| \\
&\leq \delta(|xz| + |zA|) + 2|xz| = \delta \cdot |zA| + (2+\delta)|xz| \\
&\leq \frac{1}{4}|zA| + \frac{9}{4} \cdot \frac{1}{4} \cdot |zA| \leq |zA| \\
&\leq |xA| + |xz| \leq |\tilde{x}p| + |xz|
\end{aligned}$$

which yields that $|\tilde{x}z| \leq |\tilde{x}p|$. Together with the analog estimate $|\tilde{y}z| \leq |\tilde{y}q|$ we have that

$$|\tilde{x}\tilde{y}| \leq |\tilde{x}z| + |\tilde{y}z| \leq |\tilde{x}p| + |\tilde{y}q|.$$

Thus we can consider the triangle $p'\tilde{x}'\tilde{y}'$ in $\mathbb{R}^2$ satisfying $|p'\tilde{x}'| = |p\tilde{x}|$, $|p'\tilde{y}'| = |q\tilde{y}|$, $|\tilde{x}'\tilde{y}'| = |\tilde{x}\tilde{y}|$. Let $x' \in p'\tilde{x}'$ and $y' \in p'\tilde{y}'$ be the corresponding points to x and y, respectively. The mapping $p \mapsto p'$, $q \mapsto p'$, $\tilde{x} \mapsto \tilde{x}'$, $\tilde{y} \mapsto \tilde{y}'$ is 1-Lipschitz and extends by Theorem 1.5. Hence we have that $|xy| \geq |x'y'|$ and therefore

$$|\tilde{x}\tilde{y}| = |\tilde{x}'\tilde{y}'| = (1+\delta)|x'y'| \leq (1+\delta)|xy|.$$

2.3. Boundary strata and doubling

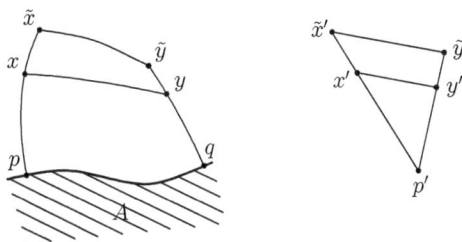

This proves that E^δ is locally Lipschitz, and the Lipschitz constant is always $1+\delta$. Now let $R > 0$ and $B(A,R) := \{x \in M \mid 0 < |xA| < R\}$. The map

$$E^\delta\big|_{W^\delta \cap B(A,R)} : W^\delta \cap B(A,R) \to B(A,(1+\delta)R)$$

is surjective and locally $(1+\delta)$-Lipschitz. Since M is second countable, and so is $W^\delta \cap B(A,R)$, we obtain that

$$\mu_n(B(A,(1+\delta)R)) \leq (1+\delta)^n \mu_n(W^\delta \cap B(A,R)).$$

Now assume that A is compact and therefore $\mu_n(B(A,R)) < \infty$ (by the Bishop-Gromov inequality). Then we let $\delta \searrow 0$ and obtain that

$$\mu_n(B(A,R)) \leq \mu_n\left(\bigcup_{0<\delta\leq 1/4} (W^\delta \cap B(A,R))\right) = \mu_n(B(A,R) \setminus X)$$

for each $R > 0$. Therefore, we have that $\mu_n(X) = 0$.

If A is not compact, let A_i be a sequence of compact sets satisfying $A_i \subseteq A_{i+1}$ and $\bigcup_{i \in \mathbb{N}} A_i = A$. For the corresponding sets X_i we have that $\mu_n(X_i) = 0 \ \forall i \in \mathbb{N}$. Now the result follows from the fact that $X \subseteq \bigcup_{i \in \mathbb{N}} X_i$. □

In particular, for almost all points $x \in M \setminus A$ a shortest path from x to A is unique. This can also be obtained by Rademacher's Theorem 1.15 on page 15.

2.3 Boundary strata and doubling

For $M \in \text{ALEX}^n(\kappa)$ the boundary ∂M is the union of all primitive extremal subsets of codimension 1, compare Corollary 1.31 on page 21. In the subsequent work we deal with components of the boundary which are not necessarily primitive, but have locally constant dimension and hence are MCS-spaces. The following definition is for the present work and not standard.

2.3 Definition. Let $M \in \text{Alex}^n(\kappa)$ with $\partial M \neq \emptyset$. A union of primitive extremal subsets of codimension 1 is called a *boundary stratum*. Furthermore, a *stratification of ∂M* is a collection of boundary strata such that each primitive extremal subset of codimension 1 is contained in exactly one boundary stratum.

Thus, a boundary stratum is an MCS-space of codimension 1, while the intersection of two distinct elements of a stratification of ∂M (this intersection is an extremal subset according to Theorem 1.25 on page 18) has codimension ≥ 2. Trivial examples of stratifications of ∂M are, of course, the collection of primitive extremal subsets of codimension 1, or just the entire boundary itself.

If $M \in \text{Alex}^n(\kappa)$ has nonempty boundary, two isometric copies of M can be glued together along their boundaries. This yields the so-called *doubling* $\bar{M} \in \text{Alex}^n(\kappa)$ with empty boundary. The procedure was developed by Perelman in [Per 91, Theorem 5.2] (Doubling Theorem) and generalized by Petrunin in [Pet 97, Theorem 2.1] (Gluing Theorem) to the case that two Alexandrov spaces with isometric boundaries are glued together. For the present work we need a slight extension of the Doubling Theorem, such that only some boundary strata are glued together with their counterparts.

2.4 Definition. For $M \in \text{Alex}^n(\kappa)$ with $\partial M \neq \emptyset$ let F be a boundary stratum. Let M_1 be isometric to M via $\varphi \colon M \to M_1$; in particular, $\varphi(F) \subseteq M_1$ is a boundary stratum of M_1. Then the *doubling* $\bar{M}$ is the metric space obtained by gluing M and M_1 via $\varphi\big|_F$. More precisely, $\bar{M} = M \sqcup M_1 / \sim$ where $M \sqcup M_1$ denotes the disjoint union and $p \sim q :\Longleftrightarrow p \in F, q = \varphi(q)$ or $q \in F, p = \varphi(q)$. The metric $d_{\bar{M}}$ on $\bar{M}$ is given as follows (we identify F with $M \cap M_1$):

$$d_{\bar{M}}(p,q) = \begin{cases} |pq| & : p,q \in M \text{ or } p,q \in M_1 \\ \min_{r \in F}\left(|pr| + |qr|\right) & : p \in M,\, q \in M_1 \end{cases}$$

2.5 Theorem (Doubling Theorem). *Let $M \in \text{Alex}^n(\kappa)$ with $\partial M \neq \emptyset$ and let F be a boundary stratum. Then $\bar{M}$ is an Alexandrov space $\bar{M} \in \text{Alex}^n(\kappa)$.*

The proof is almost the same as Perelman's. We include it here, since the paper [Per 91] is not published. A proof will also be included in the upcoming book on Alexandrov geometry by Alexander, Kapovitch and Petrunin. The following simple lemma is for later use, too.

2.6 Lemma. *Let $M \in \text{Alex}^n(\kappa)$ and $E \subseteq M$ an extremal subset with locally constant dimension (i. e. an MCS-space). Then for all $p \in E$ the subset $\Sigma_p E \subseteq \Sigma_p$ is extremal with locally constant dimension. In particular, if E is a boundary stratum of M, then $\Sigma_p E$ is a boundary stratum of Σ_p.*

2.3. Boundary strata and doubling

Proof. Let $\dim E = k$ and $p \in E$. The subset $\Sigma_p E$ is extremal according to Theorem 1.25 on page 18. An open subset $U \mathring{\subseteq} \Sigma_p E$ induces an open subset of $K(\Sigma_p E) \setminus \{\text{apex}\}$. Since the cone $K(\Sigma_p E)$ is pointed homeomorphic to some open neighborhood of p in E and E has everywhere local dimension k, the set U has dimension $k - 1$. □

Proof of the Doubling Theorem. First of all, $\bar{M}$ is again a complete length space. This follows immediately (note that F is closed); see also [BBI01, Section 3.1] for gluing procedures of length spaces.

Now we argue by induction on $n = \dim M$, the base step $n = 1$ being trivial. Since F is an extremal subset, any shortest path in M with an interior point lying in F has to lie entirely in F. Therefore, if two points $p, q \in \bar{M}$ do not lie both in F, each shortest path pq can intersect F in at most one point. Otherwise, the canonical projection of pq onto M would contain a shortest path in M with an interior point in F.

Let $p \in F$ and $a \in M$, $a_1 \in M_1$ be distinct from p. We fix shortest paths $pa \subseteq M$ and $pa_1 \subseteq M_1$. According to Lemma 2.6, the subset $\Sigma_p F$ is a boundary stratum of Σ_p and is not empty if $n \geq 2$. Hence, by induction assumption, we have that $\bar{\Sigma}_p \in \text{ALEX}^{n-1}(1)$. We want to prove the following assertion:

$$\angle a p a_1 = |\uparrow_p^a \uparrow_p^{a_1}| \tag{1}$$

The directions on the right hand side satisfy $\uparrow_p^a, \uparrow_p^{a_1} \in \bar{\Sigma}_p$. The angle on the left hand side is defined as the lower angle in length spaces, compare [BBI01, Section 3.6.5]:

$$\angle a p a_1 := \liminf_{\substack{x, x_1 \to p \\ x \in pa, x_1 \in pa_1}} \tilde{\angle}_\kappa x p x_1$$

If $\uparrow_p^a \in \Sigma_p F$ or $\uparrow_p^{a_1} \in \Sigma_p F$, there is nothing to prove in (1). Thus, we may assume that $a, a_1 \notin F$. Let $x \in pa$ and $x_1 \in pa_1$ be distinct from p and set $y := xx_1 \cap F$. By triangle comparison we obtain that

$$\tilde{\angle}_\kappa x p x_1 \geq \tilde{\angle}_\kappa x p y + \tilde{\angle}_\kappa y p x_1.$$

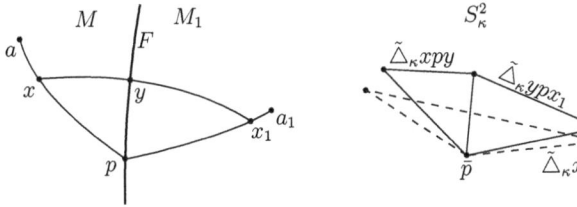

Furthermore, there is some $\nu > 0$ and $\xi \in \Sigma_p F$ such that

$$\tilde{\angle}_\kappa xpy + \tilde{\angle}_\kappa ypx_1 \geq \angle xpy + \angle ypx - 2\nu \geq |\uparrow_p^x \xi| + |\uparrow_p^{x_1} \xi| - 4\nu$$

holds, where ν can be made arbitrary small by choosing x and x_1 sufficiently close to p. This is a consequence of Theorem 1.6 on page 11 (compare also [BBI 01, Remark 10.9.4]) and $\Sigma_p F$ being extremal (compare [PP 94, Proposition 3.3]). Therefore we have that

$$\angle apa_1 \geq |\uparrow_p^a \uparrow_p^{a_1}|.$$

In particular, in the case $|\uparrow_p^a \uparrow_p^{a_1}| = \pi$ the assertion (1) is proved. Therefore, we assume that $|\uparrow_p^a \uparrow_p^{a_1}| < \pi$. Let $\nu > 0$ be given small enough to fulfill $|\uparrow_p^a \uparrow_p^{a_1}| < \pi - 4\nu$. Now choose $\xi \in \Sigma_p F$ such that $|\uparrow_p^a \xi| + |\uparrow_p^{a_1} \xi| \leq |\uparrow_p^a \uparrow_p^{a_1}| + \nu$ and choose $y \in F$ distinct from p such that $|\uparrow_p^y \xi| \leq \nu$. If y is close enough to p (which we may assume), it is possible to choose points $x \in pa$, $x_1 \in pa_1$ distinct from p such that the inequality

$$\tilde{\angle}_\kappa xpx_1 \leq \tilde{\angle}_\kappa xpy + \tilde{\angle}_\kappa ypx_1$$

holds. Then we obtain that

$$\tilde{\angle}_\kappa xpy + \tilde{\angle}_\kappa ypx_1 \leq \angle xpy + \angle ypx_1 \leq |\uparrow_p^a \uparrow_p^y| + |\uparrow_p^{a_1} \uparrow_p^y| \leq |\uparrow_p^a \xi| + |\uparrow_p^{a_1} \xi| + 2\nu \leq |\uparrow_p^a \uparrow_p^{a_1}| + 3\nu$$

which yields

$$\angle apa_1 \leq |\uparrow_p^a \uparrow_p^{a_1}|.$$

Thus, (1) is proved. In particular, if apa_1 is a shortest path, we have that $|\uparrow_p^a \uparrow_p^{a_1}| = \pi$. Furthermore, since $\bar{\Sigma}_p \in \text{ALEX}^{n-1}(1)$, we obtain that

$$|\uparrow_p^a \xi| + |\uparrow_p^{a_1} \xi| \leq \pi \quad \forall \xi \in \bar{\Sigma}_p. \tag{2}$$

This implies that for $b = b_1 \in F$ and symmetric shortest paths $pb \subseteq M$, $pb_1 \subseteq M_1$ the following holds:

$$|\uparrow_p^a \uparrow_p^b| + |\uparrow_p^{a_1} \uparrow_p^{b_1}| \leq |\uparrow_p^a \uparrow_p^b| + |\uparrow_p^{a_1} \uparrow_p^b| \leq \pi \tag{3}$$

Now the angle comparison condition can be proved, first for some special triangles. Let $\triangle baa_1 \subseteq \bar{M}$ with $b \in F$, $a \in M \setminus F$, $a_1 \in M_1 \setminus F$ and set $p := aa_1 \cap F$. Choose symmetric shortest paths $pb \subseteq M$, $pb_1 \subseteq M_1$. By (3) we have that

$$\tilde{\angle}_\kappa bpa + \tilde{\angle}_\kappa bpa_1 \leq |\uparrow_p^{a_1} \uparrow_p^{b_1}| + |\uparrow_p^a \uparrow_p^b| \leq \pi$$

2.3. Boundary strata and doubling

and can apply Alexandrov's Lemma, see [BBI 01, Lemma 4.3.3]. We obtain that

$$\tilde{\measuredangle}_\kappa baa_1 \leq \tilde{\measuredangle}_\kappa bap \leq \measuredangle bap = \measuredangle baa_1$$

and analogously

$$\tilde{\measuredangle}_\kappa ba_1 a \leq \tilde{\measuredangle}_\kappa ba_1 p \leq \measuredangle ba_1 p = \measuredangle ba_1 a.$$

For $t \geq 0$ small enough let $a(t) \in ab$ such that $|a\,a(t)| = t$. As t increases from 0, the distance $|a_1 a(t)|$ is in first order bounded below by the corresponding distance in $\tilde{\triangle}_\kappa aba_1$, since $\measuredangle baa_1 \geq \tilde{\measuredangle}_\kappa ba_1 a$. Therefore, we obtain that

$$\liminf_{t \to 0+} \frac{\tilde{\measuredangle}_\kappa a_1 ba(t) - \tilde{\measuredangle}_\kappa a_1 ba}{t} \geq 0.$$

This is true for any such triangle with the point b fixed and hence implies monotonicity of angles at b. In particular, $\measuredangle aba_1$ exists and satisfies $\tilde{\measuredangle}_\kappa aba_1 \leq \measuredangle aba_1$. Moreover, it follows from (1) that the space of directions $\Sigma_p \bar{M}$ exists and coincides with $\bar{\Sigma}_p$.

Finally, the angle comparison condition holds for any triangle in $\bar{M}$. Let $\triangle abc \subseteq \bar{M}$ with $a, b \in M \setminus F$, $c \in M_1 \setminus F$ and let $p := ac \cap F$. According to (2) we have that

$$\tilde{\measuredangle}_\kappa apb + \tilde{\measuredangle}_\kappa bpc \leq |\uparrow_p^a \uparrow_p^b| + |\uparrow_p^b \uparrow_p^c| \leq \pi$$

and can again apply Alexandrov's Lemma. This implies angle comparisons in an analogous way as above. In addition, adjacent angles always sum up to π, which completes the proof of the angle condition as given in [BBI 01, Definition 4.1.15] □

If this version of the Doubling Theorem is applied to the boundary stratification consisting of just the boundary itself, we get the standard version of the Theorem. In particular, $\bar{M}$ has empty boundary in this case. Otherwise we observe the following fact.

2.7 Corollary. *Let $M \in \text{ALEX}^n(k)$ and E, F be boundary strata. Let $\bar{M}$ be the doubling obtained by gluing along E. Then $\bar{F}$ (i.e. the set of points in $\bar{M}$ whose canonical projection onto M lies in F) is a boundary stratum of $\bar{M}$.*

This is an immediate consequence of the definitions and the doubling procedure.

2.8 Corollary. *Let $M \in \text{ALEX}^n(\kappa)$ with $\kappa \geq 0$ and let F be some boundary stratum. Then the distance function $d_F := d(F, \cdot)$ is concave on $M \setminus F$. It is strictly concave if $\kappa > 0$.*

Proof. According to [Per 91, Theorem 6.1] the distance function $d_{\partial M}$ to the boundary is concave on $M \setminus \partial M$ (strictly if $\kappa > 0$); see also [Pet 07, Theorem 3.3.1]. Thus, the assertion is proved if $F = \partial M$. Otherwise let $E := \overline{\partial M \setminus F}$. Clearly, E is a boundary stratum. If we glue along E and consider the doubling $\bar{M}$, we have that $\partial \bar{M} = \bar{F}$. Therefore, the distance function $d_{\bar{F}}$ is (strictly) concave on $\bar{M} \setminus \bar{F}$, which yields that d_F is (strictly) concave on $M \setminus F$. □

It is of course possible to prove the statement directly by modifying the proofs of Perelman or Petrunin, respectively.

Note that in fact d_F restricted to any shortest path $pq \subseteq M$ is a concave function, also if $pq \cap F \neq \emptyset$. This is clear by continuity, as well as the fact that "strictly" does not carry over. We will often use this concavity on the entire space M, but strictly speaking d_F on M is not (semi-) concave in the sense of Definition 1.18 on page 16. In particular, the gradient $\nabla_p d_F$ is only defined for $p \in M \setminus F$, so far. By the next lemma we can remove this deficiency.

2.9 Lemma. *For $M \in \text{ALEX}^n(0)$ let F be some boundary stratum and $p \in F$. Then there exists a unique direction $\xi \in \Sigma_p$ where d_F increases most. Moreover, if E is another boundary stratum and $p \in E$, then $\xi \in \Sigma_p E$.*

Proof. First of all, there are directions in which d_F increases, namely $\uparrow_p^q$ for each shortest path pq with $q \in M \setminus F$. Take $\xi \in \Sigma_p$ such that $d_{\Sigma_p F}$ attains its maximum at ξ. According to Corollary 2.8 this maximum point is unique since $d_{\Sigma_p F}$ is strictly concave on $\Sigma_p \setminus \Sigma_p F$.

If E is another boundary stratum with $p \in E$, we glue along E and consider the doubling $\bar{M}$ and $\bar{\Sigma}_p$, respectively. It is clear by the doubling definition that $\xi \in \bar{\Sigma}_p$ is also a maximum point for $d_{\bar{\Sigma}_p \bar{F}}$, and the same holds for its reflexion point. On the other hand, the procedure from above applied to the function $d_{\bar{\Sigma}_p \bar{F}}$ on $\bar{\Sigma}_p$ gives a unique maximum point. Thus, ξ and its reflexion point must coincide, which is equivalent to $\xi \in \Sigma_p E$. □

By this result, the gradient of d_F is well-defined also at $p \in F$; note that the directional derivatives exist by concavity of d_F and satisfy $d_p d_F(\xi) = -\cos\left(d_{\Sigma_p F}(\xi) + \frac{\pi}{2}\right) \quad \forall \xi \in \Sigma_p$. Lemma 1.20 on page 16 carries over and ensures the existence of unique gradient curves starting at all points in F. In addition, once some gradient curve of d_F reaches another boundary stratum E, it stays in E. The same is true, by definition, for each extremal subset not intersecting F. In fact, we have the following stronger statement.

2.10 Lemma. *For $M \in \text{ALEX}^n(0)$ let F be some boundary stratum. Let $E \subseteq M$ be a compact extremal subset satisfying $E \cap F = \emptyset$. Then d_F is constant on E, attaining its maximum.*

Proof. Since F is closed and E compact, there are points $p \in E$, $q \in F$ satisfying $|pq| = |EF| \neq 0$. Assume, by way of contradiction, that there exists $p' \in M$ such that $d_F(p') > d_F(p)$. By concavity of d_F we have that

$$0 < d_p d_F(\uparrow_p^{p'}) = -\cos |\Uparrow_p^F \uparrow_p^{p'}|$$

2.4. Superlevel sets

and therefore $\angle qpp' > \frac{\pi}{2}$. On the other hand, the distance function $d(q, \cdot)\big|_E$ attains its minimum at p. Since E is extremal, it follows that $\angle qpp' \leq \frac{\pi}{2}$, a contradiction. $\square$

As another consequence of Lemma 2.9 we formulate the following statement.

2.11 Lemma. *For $M \in \text{A\textsc{lex}}^n(0)$ let F be some boundary stratum and $p \in M \setminus F$. Let $q \in F$ be the endpoint of some shortest path from p to F. Then qp is perpendicular to F, i. e. $|\uparrow_q^p \xi| = \frac{\pi}{2} \ \forall \xi \in \Sigma_q F$. Moreover, if q is also the endpoint of some shortest path from $p' \in M \setminus F$ to F, then $pq \subseteq p'q$ or $p'q \subseteq pq$. In addition, if E is another boundary stratum, then $q \in E$ implies $p \in E$.*

Proof. Let $\xi \in \Sigma_q F$. We have by assumption that $|pq| = |pF|$, which implies that $|\uparrow_q^p \xi| \geq \frac{\pi}{2}$. The reverse inequality follows since F is extremal. The shortest path qp coincides with the gradient curve α_q of d_F. Therefore, it is unique in the sense of the second assertion of the Lemma. In addition, if $q \in E$, then Lemma 2.9 implies that $\alpha_q \subseteq E$, in particular $p \in E$. $\square$

The following result is also an application of concavity and extremality.

2.12 Lemma. *Let $M \in \text{A\textsc{lex}}^n(0)$ and let F_1, F_2 be two elements of some stratification of ∂M. Then for any point $p \in M \setminus (F_1 \cup F_2)$ we have that*

$$|\Uparrow_p^{F_1} \Uparrow_p^{F_2}| \geq \frac{\pi}{2}.$$

Proof. Let $q_i \in F_i$, $i = 1, 2$ such that $|pq_i| = |pF_i|$. Assume, by way of contradiction, that $\angle q_1 p q_2 < \frac{\pi}{2}$ and hence $d_p d_{F_1}(\uparrow_p^{q_2}) < 0$. Let $r \in F_1$ be closest to q_2, then concavity of d_{F_1} implies that $d_p d_{F_1}(\uparrow_{q_2}^p) > 0$ and therefore $\angle pq_2 r > \frac{\pi}{2}$. This contradicts the extremality of F_2. $\square$

We conclude this section with the remark that for d_{F_i} each local maximum is in fact global. This follows immediately from concavity.

2.4 Superlevel sets

The fact that the distance function to the boundary of an Alexandrov space $M \in \text{A\textsc{lex}}^n(0)$ is concave implies that superlevel sets of this function are convex subsets. In particular, if M is compact, the (super)level set to the maximal value of $d_{\partial F}$ is an Alexandrov space of dimension $< n$. If it has nonempty boundary, the procedure can be iterated and gives in the end a convex subset $S \subseteq M$ without boundary, called the *soul* of M. This approach was developed by Perelman in [Per 91, Section 6] in order to generalize the Soul Theorem by Cheeger and Gromoll (see [CG 72]) to Alexandrov spaces. More precisely, Perelman proved that the procedure described above also works in the non-compact case and that the resulting soul is a deformation retract of M.

If $M \in \text{ALEX}^n(0)$ is compact and a metric product $M = A \times B$ of Alexandrov spaces A, B with nonnegative curvature, each factor, say A, can be considered as a "soul" of M (it might have boundary, though). Moreover, the entire space M is fibrated into isometric copies of A. Thus, a basic step in the proof of the Splitting Theorem is to find a fibration of M into convex subsets which are all isometric.

In order to find some "soul" of $M \in \text{ALEX}^n(0)$ at each point $p \in M$, we consider superlevel sets of distance functions d_F, where F is some boundary stratum. The (super)level set to the maximum value of d_F is a convex set. Now the question arises how to construct such convex sets at each point in M. The principle is the following: If E is another boundary stratum, it turns out that each superlevel set M' of d_E is a space which looks essentially the same as M, as long as no collapse[1] happens. This means, we consider $M' = d_E^{-1}([s, a])$ with $a = \max_M d_E$ and $s < a$. Then $F' := F \cap M'$ is a boundary stratum in M', and the construction from above can be performed in M' giving another convex subset.

The procedure will be formalized in the next chapter. Here we prepare the basic tools.

2.13 Lemma. *Let $M \in \text{ALEX}^n(0)$ be compact and let F_1 be some boundary stratum. Then $M' := d_{F_1}^{-1}([s, a_1])$ satisfies $M' \in \text{ALEX}^n(0)$ for all $s < a_1 := \max_{p \in M} d_{F_1}(p)$. In addition, if F_2 is another boundary stratum, we set $F_2' := F_2 \cap M'$. Then for the space M' the following holds: $d_{F_2'} = d_{F_2}\big|_{M'}$ and F_2' is a boundary stratum of M'.*

Proof. The function d_{F_1} is concave and therefore, $M' \subseteq M$ is a convex subset, in particular an Alexandrov space of nonnegative curvature. Since the Hausdorff dimension of M is locally constant and $s < a_1$ by assumption, we have that $\dim_H M' = \dim_H M = n$.

Let $p \in M' \setminus F_2'$ and $q \in F_2$ such that $d_{F_2}(p) = |pq|$. We claim that this implies $q \in F_2'$ and hence $d_{F_2}\big|_{M'} = d_{F_2'}$. First, we may assume that $q \notin F_1$, because by Lemma 2.11 on the preceding page, $q \in F_1$ implies $p \in F_1$ and therefore the trivial case that $M' = M$. Take $r \in F_1$ at minimal distance to $q \notin F_1$. By extremality of F_2 and the choice of q, we obtain that $\angle pqr \leq \frac{\pi}{2}$ and hence $d_q d_{F_1}(\uparrow_q^p) \leq 0$. Concavity of d_{F_1} implies that $d_{F_1}(q) \geq d_{F_1}(p) \geq s$, which means that $q \in M'$ and proves the claim.

Now it is clear that the following holds: If $p \in M' \setminus F_2'$ is given and the function $d_p\big|_{F_2'}$ attains its minimum at $q \in F_2'$, we have that $\angle pqx \leq \frac{\pi}{2}$ $\forall x \in M'$. Therefore, F_2' is an extremal set in M'; recall that F_2' is closed. In addition, we have that $\overline{F_2' \setminus d_{F_1}^{-1}(s)} = F_2'$. Indeed, for $x \in F_2' \cap d_{F_1}^{-1}(s)$ the gradient curve to d_{F_1} stays in F_2' and immediately leaves the level

---

[1]in other words, as long as $\dim_H M' = \dim_H M$. Recall that the dimension of Alexandrov spaces cannot explode under Gromov-Hausdorff limits by Gromov's Compactness Theorem.

2.4. Superlevel sets

set $d_{F_1}^{-1}(s)$. Since F_2 is a boundary stratum and $M' \subseteq M$ is a convex subset, the set of points $x \in F_2' \setminus d_{F_1}^{-1}(s)$ with tangent cones $T_x M \approx \mathbb{R}^{n-1} \times \mathbb{R}_+$ ($\approx$ meaning homeomorphic) is dense in $F_2' \setminus d_{F_1}^{-1}(s)$, compare Corollary 1.31 on page 21. Thus, this set is also dense in F_2' and since $F_2' \subseteq M'$ is extremal, it is a boundary stratum of M'. □

Based on this result it can be shown that any stratification of ∂M gives an analog stratification of $\partial M'$.

2.14 Proposition. *Let $M \in \text{ALEX}^n(0)$ be compact and let $F_1, \ldots, F_\ell$ be some stratification of ∂M. Let $M' := d_{F_1}^{-1}([s, a_1])$ with $0 \leq s < a_1 := \max_{p \in M} d_{F_1}(p)$ and define $F_1' := d_{F_1}^{-1}(s)$ and $F_i' := F_i \cap M'$ for $i = 2, \ldots, \ell$. Then $F_1', \ldots, F_\ell'$ is a stratification of $\partial M'$ and the following holds:*

$$\bigcap_{i \in I} F_i = \emptyset \iff \bigcap_{i \in I} F_i' = \emptyset \qquad \text{for each subset } I \subseteq \{1, \ldots, \ell\}$$

Proof. According to Lemma 2.13 the sets $F_2', \ldots F_\ell'$ are boundary strata. We consider F_1' and again, for $s = 0$ there is nothing to prove. Let $A_1 := \{p \in M \mid d_{F_1}(p) = a_1\}$.

The function d_{F_1} is admissible and regular on $M \setminus (F_1 \cup A_1)$, compare Section 1.6. Hence, by the Local Fibration Theorem 1.30 on page 21, the set $F_1' = d_{F_1}^{-1}(s)$ is an MCS-space of dimension $n-1$. In particular, the set of points possessing a neighborhood $U \overset{\circ}{\subseteq} F_1'$ homeomorphic to $\mathbb{R}^{n-1}$ is dense in F_1'. Let $p \in U \overset{\circ}{\subseteq} F_1'$ be such a point. Again by the Fibration Theorem, p has some neighborhood $V \overset{\circ}{\subseteq} M$ satisfying $V \approx U \times \mathbb{R}$. This implies that $V \cap M'$ is a neighborhood of p in M' fulfilling $V \cap M' \approx U \times \mathbb{R}_+ \approx \mathbb{R}^{n-1} \times \mathbb{R}_+$. Therefore, we have that $p \in \partial M'$. Since such points p are dense in F_1' and $\partial M'$ is closed, we obtain that $F_1' \subseteq \partial M'$. This in turn implies that $F_1' \cup \ldots \cup F_\ell' \subseteq \partial M'$, and the reverse inclusion is trivial, because any boundary point in $M' \setminus (F_1' \cup \ldots \cup F_\ell')$ would be a boundary point in $M \setminus (F_1 \cup \ldots \cup F_\ell) = \emptyset$.

This implies that $\overline{\partial M' \setminus (F_2' \cup \ldots \cup F_\ell')} = \overline{F_1' \setminus (F_2' \cup \ldots \cup F_\ell')} = F_1'$ is a boundary stratum, since $F_2', \ldots, F_\ell'$ are boundary strata and the second equality follows like in the proof of Lemma 2.13. Moreover, the collection $F_1', \ldots, F_\ell'$ is a stratification of $\partial M'$.

It remains to examine the intersections of boundary strata. The stratum F_1 plays a special role, so let $I \subseteq \{2, \ldots, \ell\}$ and $F := \bigcap_{i \in I} F_i$, $F' := \bigcap_{i \in I} F_i'$.

If $F = \emptyset$, clearly also $F' = \emptyset$. Thus, let $F \neq \emptyset$ and consider the following two cases.

If $F_1 \cap F = \emptyset$, we can apply Lemma 2.10 on page 32, because F is an extremal set. This gives that $F \subseteq A_1 \subseteq M'$ and therefore $F' = F$.

If $F_1 \cap F \neq \emptyset$, let $p \in F_1 \cap F$ and consider the gradient curve α_p of the function d_{F_1}. Since $\alpha_p(t) \in F \quad \forall t \geq 0$ (by Lemma 2.9 on page 32), there exists $T > 0$ such that $d_{F_1}(\alpha_p(T)) = s$. This implies that $\alpha_p(T) \in F_1' \cap F' \neq \emptyset$ and completes the proof. □

35

Chapter 2. Tools for the Splitting Theorem

2.15 Corollary. *For some compact $M \in \text{ALEX}^n(0)$ let $F_1, \ldots, F_\ell$ be a stratification of ∂M and $0 \leq s_i < a_i := \max\limits_{p \in M} d_{F_i}(p)$ for $i = 1, \ldots, \ell$. If the space $M' := d_{F_1}^{-1}([s_1, a_1]) \cap \ldots \cap d_{F_\ell}^{-1}([s_\ell, a_\ell])$ is non-collapsed, then its boundary possesses the stratification $\partial M' = F_1' \cup \ldots \cup F_\ell'$ with boundary strata $F_i' := d_{F_i}^{-1}(s_i) \cap M'$. Moreover, the intersection of any collection of the strata F_i is empty if and only if the intersection of the corresponding strata F_i' is empty.*

Proof. This follows by iterated use of Proposition 2.14 and the fact that the order of taking superlevel sets is irrelevant according to Lemma 2.13. Indeed, in any space $M' = d_{F_i}^{-1}([s_i, a_i])$ the distance function $d_{F_j'}$ is just the restriction $d_{F_j}\big|_{M'}$ for $j \neq i$. In addition, of course, $d_{F_i'} = d_{F_i}\big|_{M'} - s_i$. □

An operation very often used in the subsequent work is the following: If $p \in M$ and F is some boundary stratum, we consider the superlevel set $M' = d_F^{-1}([d_F(p), \max d_F])$. We simply say that F *is shifted to* p and often rename M' and F' back to M and F to simplify notation.

2.5 Structure results on intersections of boundary strata

As we have seen in the last section, if M is given with some assumptions about boundary strata and their intersecting behaviour, then any superlevel set fulfills the same assumptions (as long as no collapse happens). This implies that structure results about boundary strata and in particular their intersections carry over to such superlevel sets $M' \subseteq M$. Also vice versa, some results can be obtained only for superlevel sets $M' \subsetneq M$ (since an ambient space is needed) and then may carry over to M by the Stability Theorem.

We will see later how this works in detail.

2.16 Lemma. *Let $M \in \text{ALEX}^n(0)$ be compact and let $F_1, \ldots, F_{k+1}$ be a subcollection of some stratification of ∂M. As always we set $a_i = \max\limits_{p \in M} d_{F_i}(p)$ and $A_i := \{p \in M \mid d_{F_i}(p) = a_i\}$. Assume that the following holds.*

(i) $F_1 \cap \ldots \cap F_{k+1} = \emptyset$;

(ii) $F_1 \cap \ldots \cap \widehat{F_j} \cap \ldots \cap F_{k+1} \neq \emptyset \quad \forall j \in \{1, \ldots, k+1\}$.

For $j \in \{1, \ldots, k+1\}$ let $p \in A_j$ and $q \in M$ such that for all $i \neq j$ we have that $d_{F_i}(q) < d_{F_i}(p)$ if $d_{F_i}(p) > 0$ and $d_{F_i}(q) = 0$ if $d_{F_i}(p) = 0$.

Then also $q \in A_j$. Moreover, it follows that $A_j \subseteq F_1 \cup \ldots \cup \widehat{F_j} \cup \ldots \cup F_{k+1}$.

Note that, according to a standard definition, $\widehat{X}$ means that X is omitted in the respective collection.

2.5. Structure results on intersections of boundary strata

Proof. We show the assertion for A_{k+1}. First of all, by the assumptions and Lemma 2.10 on page 32, we have that $F_1 \cap \ldots \cap F_k \subseteq A_{k+1}$. According to Corollary 2.15 on the facing page, the analog is true for each superlevel set as long as no collapse occurs.

For $i = 1, \ldots, k$ we set $s_i := d_{F_i}(p)$ and $t_i := d_{F_i}(q)$. Now consider the space $M' := d_{F_1}^{-1}([t_1, a_1]) \cap \ldots \cap d_{F_k}^{-1}([t_k, a_k])$, which is not collapsed since $t_i < s_i$ or $t_i = s_i = 0$. We have that $q \in F_1' \cap \ldots \cap F_k'$ and again by Lemma 2.10, the function $d_{F_{k+1}}\big|_{M'}$ attains its maximum at q. On the other hand we have that $p \in M'$ and therefore $a_{k+1} = d_{F_{k+1}}(p) \leq d_{F_{k+1}}(q) \leq a_{k+1}$. This implies everywhere equality and hence $q \in A_{k+1}$.

Assume now, by way of contradiction, that $s_i > 0 \ \forall i \in \{1, \ldots, k\}$. Let $z \in F_1 \cap \ldots \cap F_k$ and $r > 0$ satisfying $r < s_i \ \forall i \in \{1, \ldots, k\}$. Then, by the first part, the ball $B_r(z)$ (taken in M) is contained in A_{k+1}, a contradiction to the definition of A_{k+1}.

Hence, for each $p \in A_{k+1}$ there exists some index i such that $d_{F_i}(p) = 0$. In other words, $A_{k+1} \subseteq F_1 \cup \ldots \cup F_k$. □

The next lemma is purely technical and only a tool for the proof of the subsequent theorem.

2.17 Lemma. *Let $M \in \text{ALEX}^n(0)$ be compact and let $F_1, \ldots, F_{k+1}$ be a subcollection of some stratification of ∂M. Then the following constellation is impossible: All intersections of $k - 1$ boundary strata F_i are nonempty, while all intersections of k boundary strata are empty.*

Proof. Assume, by way of contradiction, there is such constellation. Note that only for $k \geq 2$ there is something to prove.

Let A_{k+1} as before and $G_i := F_1 \cap \ldots \cap \widehat{F_i} \cap \ldots \cap F_k$ for $i \in \{1, \ldots, k\}$. By assumption there are points $p_i \in G_i$ for each $i \in \{1, \ldots, k\}$ and these points are pairwise distinct. In addition, we have that $G_i \cap F_{k+1} = \emptyset$ and therefore by Lemma 2.10 that $G_i \subseteq A_{k+1} \ \forall i \in \{1, \ldots, k\}$. Moreover, by applying Lemma 2.16 to the subcollection $F_2, \ldots, F_{k+1}$, we obtain that $A_{k+1} \subseteq F_2 \cup \ldots \cup F_k$.

Let $q_1 \in p_1 p_2$ be some interior point. Since A_{k+1} is convex, we have that $q_1 \in A_{k+1}$. The fact that $p_1 \notin F_1$ and $p_2 \notin F_2$ implies that $q_1 \in (F_2 \cup \ldots \cup F_k) \setminus (F_1 \cup F_2)$, because all F_i are extremal. Now let $q_2 \in q_1 p_3$ be some interior point and obtain that $q_2 \in (F_2 \cup \ldots \cup F_k) \setminus (F_1 \cup F_2 \cup F_3)$ and iterate. This process leads to some point $q_{k-1} \in q_{k-2} p_k$ satisfying the condition $q_{k-1} \in (F_2 \cup \ldots \cup F_k) \setminus (F_1 \cup \ldots \cup F_k)$, a contradiction. □

The following theorem ensures that intersections of k boundary strata are MCS-spaces of the expected topological dimension.

2.18 Theorem. *Let $M \in \text{ALEX}^n(0)$ be compact with $\partial M \neq \emptyset$. Then the following holds for any subcollection $F_1, \ldots, F_k$ of any stratification of ∂M: If $k \leq n$ and $F_1 \cap \ldots \cap F_k \neq \emptyset$, then this intersection is an MSC-space satisfying $\dim(F_1 \cap \ldots \cap F_k) = n - k$. If $k > n$, then $F_1 \cap \ldots \cap F_k$ is always empty.*

Proof. The proof is carried out via inductions on k and on n. For $k = 1$ the statement is clearly true for all $n \in \mathbb{N}$. Now assume that the statement is proved for any intersection of k boundary strata and for all $n \in \mathbb{N}$. We will prove the statement for intersections of $k+1$ boundary strata and, by a second induction on n, for all $n \in \mathbb{N}$.

(i) If $n = 1, \ldots, k-1$, in other words if $n < k$, then by induction assumption on k, we have that any intersection of k boundary strata is empty, hence also any intersection of $k+1$ boundary strata.

(ii) If $n = k$, the case $n = 1$ is already proved. Let $n \geq 2$ and assume, by way of contradiction, there exists $p \in F_1 \cap \ldots \cap F_{k+1}$. The space of directions Σ_p fulfills the following: $\dim \Sigma_p = n - 1$ and $\Sigma_p F_1, \ldots, \Sigma_p F_{k+1}$ are boundary strata according to Lemma 2.6 on page 28. Moreover, these sets form a subcollection of some stratification of $\partial \Sigma_p$, i.e. the intersection of each two boundary strata is not a boundary stratum. Indeed, if, say, $\dim(\Sigma_p F_1 \cap \Sigma_p F_2) = n - 2$, then $\dim(F_1 \cap F_2) = n - 1$, which is impossible since the sets F_i form a subcollection of a stratification of ∂M.

Now we can apply (i) and obtain that $\Sigma_p F_1 \cap \ldots \cap \widehat{\Sigma_p F_i} \cap \ldots \cap \Sigma_p F_{k+1} = \emptyset$ for all $i \in \{1, \ldots, k+1\}$. We claim that, in contrast, any intersection of $k - 1$ boundary strata $\Sigma_p F_i$ is nonempty. The claim is proved via another induction and for a fixed numbering, which is of course without loss of generality.

The base step $\Sigma_p F_1 \neq \emptyset$ holds by assumption (i.e. $n \geq 2$).

Now assume that $\Sigma_p F_1 \cap \ldots \cap \Sigma_p F_\ell \neq \emptyset$ is proved for all $\ell \leq k - 2$. By the induction assumption for k we have that $\dim(\Sigma_p F_1 \cap \ldots \cap \Sigma_p F_\ell) = (n-1) - \ell$ and in turn $(n-1) - \ell \geq (n-1) - (k-2) = (n-1) - (n-2) = 1$. Being a boundary stratum, the set $\Sigma_p F_{\ell+1}$ has codimension 1 in Σ_p. Hence, by Petrunin's version of Frankel's Theorem in [Pet98, Corollary 3.3], the following intersection is not empty: $\Sigma_p F_1 \cap \ldots \cap \Sigma_p F_\ell \cap \Sigma_p F_{\ell+1} \neq \emptyset$ and thus, the claim is proved.

Consequently, we can apply Lemma 2.17 and have a contradiction. Hence, $F_1 \cap \ldots \cap F_{k+1} = \emptyset$ is proved.

(iii) If $n = k+1$, let $p \in F_1 \cap \ldots \cap F_{k+1}$. As shown in step (ii), we can apply the induction assumption to the boundary strata $\Sigma_p F_1, \ldots, \Sigma_p F_{k+1} \subseteq \Sigma_p$. This gives that $\Sigma_p F_1 \cap \ldots \cap \Sigma_p F_{k+1} = \emptyset$, since $\dim \Sigma_p = n - 1 = k < k+1$. Therefore, we have that $\dim(F_1 \cap \ldots \cap F_{k+1}) = 0 = n - (k+1)$, which is the statement of the Theorem.

(iv) If $n > k+1$, assume that the statement is proved for dimension $n - 1$. Now let the dimension be n and consider $p \in F_1 \cap \ldots \cap F_{k+1} \neq \emptyset$. Again, in the space of directions Σ_p the induction assumption can be applied. By the fact that $\dim \Sigma_p = n - 1 \geq k+1$, the same argument as in case (ii) via Frankel's/Petrunin's Theorem shows that $\Sigma_p F_1 \cap \ldots \cap \Sigma_p F_{k+1} \neq \emptyset$.

2.5. Structure results on intersections of boundary strata

Hence, by induction assumption, we obtain that $\dim(\Sigma_p F_1 \cap \ldots \cap \Sigma_p F_{k+1}) = (n-1) - (k+1)$, which implies that $\dim(F_1 \cap \ldots \cap F_{k+1}) = n - (k+1)$ as desired. Indeed, since p was arbitrary, the dimension is locally constant. In other words, $F_1 \cap \ldots \cap F_{k+1}$ is an MCS-space.

This concludes the proof of the Theorem. □

2.19 Corollary. *For $\Sigma \in \text{ALEX}^n(1)$ let $F_1, \ldots, F_k$ be a subcollection of some stratification of $\partial \Sigma$. Then we have that $F_1 \cap \ldots \cap F_k \neq \emptyset$ if and only if $k \leq n$.*

Proof. This follows immediately from Theorem 2.18 and Frankel's/Petrunin's Theorem. □

Recall that for $M \in \text{ALEX}^n(0)$ boundary points can be recognized as follows: If $T_p M \approx \mathbb{R}^{n-1} \times \mathbb{R}_+$, then $p \in \partial M$. (In fact, ∂M is the closure of all such points.) If F is the intersection of some boundary strata, the question arises if F has "boundary". However, F is in general not an Alexandrov space, hence the term "boundary" is not defined. But since F is an MCS-space, one can at least ask if there are points $p \in F$ satisfying $T_p F \approx \mathbb{R}^{\ell-1} \times \mathbb{R}_+$, where $\ell = \dim F$.

In order to deal with this problem we want to use the Relative Local Fibration Theorem. Since it cannot be applied to boundary strata of M directly, we will work in some superlevel set Hausdorff-close to M. For this reason we need the following lemma.

2.20 Lemma. *Let $M \in \text{ALEX}^n(0)$ be compact and let $E, F_1, \ldots, F_k$ be some stratification of ∂M. We set $a := \max_{p \in M} d_E(p)$ and choose a sequence $(s_i) \subseteq [0, a)$ decreasing to 0. Let $M_i := d_E^{-1}([s_i, a])$ and $E_i := d_E^{-1}(s_i)$ and let $F = \bigcap_{\ell \in I} F_\ell$ be the intersection of boundary strata with some index set $I \subseteq \{1, \ldots, k\}$. Then we have the Hausdorff convergences $M_i \xrightarrow{i \to \infty} M$, $E_i \xrightarrow{i \to \infty} E$ and $E_i \cap F \xrightarrow{i \to \infty} E \cap F$.*

Proof. Since s_i is decreasing, we have that $M_i \subseteq M_{i+1} \; \forall i$. Thus, the convergence $M_i \to M$ is clear. Now, if $\varepsilon > 0$ is given, choose i big enough such that $\max(s_i, d_H(M_i, M)) < \varepsilon$ (where d_H denotes the Hausdorff distance). Then by definition $|pM_i| < \varepsilon \; \forall p \in M$, which implies that $|pE_i| < \varepsilon \; \forall p \in E$. On the other hand we have that $|qE| = s_i < \varepsilon \; \forall q \in E_i$ and hence, $E_i \to E$. In order to show the remaining convergence, note that according to the Compactness Theorem by Blaschke (see e.g. [BBI01, Theorem 7.3.8]), we may assume that $E_i \cap F$ has a partial limit, say L. It suffices to show that $L = E \cap F$.

The Hausdorff limit L consists precisely of all points which are limit points of sequences in $E_i \cap F$. If p_i is a sequence converging to $p \in M$ and satisfying $p_i \in E_i \cap F$, then clearly $p \in E \cap F$, since $E_i \to E$ and F is closed. This implies that $L \subseteq E \cap F$. For the other implication, let $q \in E \cap F$ and let α_q be the gradient curve of the function d_E starting at q. According to Lemma 2.9 on page 32, the gradient curve α_q stays in F. By setting $q_i := \alpha_q \cap E_i \; \forall i$ we obtain a sequence $q_i \to q$ with $q_i \in E_i \cap F \; \forall i$. Indeed, $d_E(q_i) = s_i \searrow 0$ implies that $q_i \to q$. Thus we have that $L = E \cap F$ as desired. □

Chapter 2. Tools for the Splitting Theorem

Now we come back to the question asked above.

2.21 Proposition. *Let $M \in \text{ALEX}^n(0)$ be compact and let $F_1, \ldots, F_k$ be a subcollection of some stratification of ∂M. Assume that $F := F_1 \cap \ldots \cap F_k \neq \emptyset$. Let $p \in F$ such that $T_p F \approx \mathbb{R}^{n-k-1} \times \mathbb{R}_+$.*
Then the point p is contained in some boundary stratum distinct from $F_1, \ldots, F_k$, i.e. $p \in \overline{\partial M \setminus (F_1 \cup \ldots \cup F_k)}$. In particular, if there is no additional boundary stratum intersecting with $F_1 \cap \ldots \cap F_k$, such points p do not exist.

Proof. The assertion is proved via induction on k. Let $k = 1$, i.e. F is a boundary stratum. Let $p \in F$ with $T_p F \approx \mathbb{R}^{n-2} \times \mathbb{R}_+$ and assume, by way of contradiction, that p lies in no other boundary stratum. Hence we can assume that $\partial M = F$. Indeed, if $\partial M \setminus F \neq \emptyset$, consider the doubling $\bar{M}$ obtained by gluing along the boundary stratum $\overline{\partial M \setminus F}$. This does not affect $T_p F = T_p \bar{F}$ by the assumption on p.

Let $a := \max_{x \in M} d_F(x)$, $s \in (0, a)$ and consider the level set $F' := d_F^{-1}(s)$. We claim that there is no open subset $U \overset{\circ}{\subseteq} F'$ homeomorphic to $\mathbb{R}^{n-2} \times \mathbb{R}$. Otherwise let $q \in U$. Since d_F is admissible and regular near q, we can apply the Local Fibration Theorem 1.30 on page 21. Then q has a neighborhood $V \overset{\circ}{\subseteq} M$ satisfying $V \approx U \times \mathbb{R} \approx \mathbb{R}^{n-1} \times \mathbb{R}_+$. This implies that $q \in \partial M = F$, which contradicts the definition of F'.

Now we consider F' as the boundary of the space $M' = d_F^{-1}([s, a])$ (according to Proposition 2.14 on page 35). The claim from above is true for each $s \in (0, a)$. By Lemma 2.20, any sequence $s_i \searrow 0$ induces a sequence of spaces converging to M with boundaries converging to F. According to the Relative Stability Theorem [Kap 07, Theorem 9.2], we have that $F \approx F'$ for s small enough. By the choice of p, there is some open subset in F homeomorphic to $\mathbb{R}^{n-2} \times \mathbb{R}_+$. Hence, such subset exists also in F', a contradiction to the claim. Thus, the base step of the induction is proved.

The proof of the induction step is similar. Let $p \in F = F_1 \cap \ldots \cap F_k$ such that $T_p F \approx \mathbb{R}^{n-k-1} \times \mathbb{R}_+$ and assume that p lies in no other boundary stratum. Analogously to above, we then can assume that $F_1, \ldots, F_k$ is a stratification of ∂M.

For $a_1 := \max_{x \in M} d_{F_1}(x)$ and $s \in (0, a_1)$ we consider the set $F' := d_{F_1}^{-1}(s) \cap F_2 \cap \ldots \cap F_k$. Our claim is that there is no open subset $U \overset{\circ}{\subseteq} F'$ homeomorphic to $\mathbb{R}^{n-k-1} \times \mathbb{R}_+$. Assume the contrary and let $q \in U$. The function d_{F_1} is admissible and regular near q and the set $F_2 \cap \ldots \cap F_k$ is extremal. Hence, the Relative Local Fibration Theorem can be applied, see [Kap 07, Theorem 9.7]. This gives that q has a neighborhood $V \overset{\circ}{\subseteq} F_2 \cap \ldots \cap F_k$ satisfying $V \approx U \times \mathbb{R} \approx \mathbb{R}^{n-k} \times \mathbb{R}_+$. By the induction assumption it follows that q lies in some boundary stratum distinct from $F_2, \ldots, F_k$. This implies $q \in F_1$, a contradiction to the definition of F'.

2.5. Structure results on intersections of boundary strata

Thus, the claim is proved for each $s \in (0, a_1)$. Now the contradiction to the assumption on p follows analogously to the case $k = 1$, using Lemma 2.20 and the Relative Stability Theorem. □

If the intersection $F_1 \cap \ldots \cap F_k$ is a convex set and therefore an Alexandrov space, its boundary coincides with the closure of points p as in Proposition 2.21. If $A \subsetneq F_1 \cap \ldots \cap F_k$ is a proper subset, but still of the same dimension and convex, one expects additional points of ∂A, namely all points of the topological boundary of A inside $F_1 \cap \ldots \cap F_k$. In the following we give more precise statements on this issue.

2.22 Lemma. *Let $\Sigma \in \mathrm{ALEX}^n(1)$ and let $F_1, \ldots, F_k$ be a subcollection of some stratification of $\partial \Sigma$ with $k < n$, $n \geq 2$. Then the intersection $F_1 \cap \ldots \cap F_k$ is connected.*

Proof. First, recall that $F_1 \cap \ldots \cap F_k \neq \emptyset$ by Corollary 2.19 on page 39. Now we use induction on k. The base step $k = 1$ is clear by Frankel's/Petrunin's Theorem. Indeed, $\dim F_1 = n - 1$ and $2(n-1) \geq n$ since $n \geq 2$. Therefore, two connected components of F_1 would intersect.

For the proof of the induction step let $p, q \in F_1 \cap \ldots \cap F_k$. Since d_{F_1} is strictly concave (see Corollary 2.7 on page 31), the subset in Σ where d_{F_1} attains its maximum consists of one point only. Let z_1 be this point. Moreover, Lemma 2.9 on page 32 implies that $z_1 \in F_2 \cap \ldots \cap F_k$, because all gradient curves of the function d_{F_1} end at the point z_1. In particular, there is a curve $\gamma \subseteq F_2 \cap \ldots \cap F_k$ from p to q.

We claim that γ can be chosen such that $z_1 \notin \gamma$. Assume the contrary, then for each neighborhood $U \subseteq F_2 \cap \ldots \cap F_k$ of z_1 the set $U \setminus \{z_1\}$ is not path-connected. This implies that $K(\Sigma_{z_1} F_2 \cap \ldots \cap \Sigma_{z_1} F_k) \setminus \{\mathrm{apex}\}$ is not path-connected and therefore also $\Sigma_{z_1} F_2 \cap \ldots \cap \Sigma_{z_1} F_k$ is not path-connected. By induction assumption, $\Sigma_{z_1} F_2 \cap \ldots \cap \Sigma_{z_1} F_k$ is connected and hence also path-connected, since it is a stratified manifold. This contradiction proves the claim.

Let $x \in \Sigma \setminus (F_1 \cup \{z_1\})$. The concavity of d_{F_1} implies that $|\Uparrow_x^{F_1} \Uparrow_x^{z_1}| > \frac{\pi}{2}$. Therefore, the gradient $\nabla_x f$ of the function $f := d_{z_1}^2$ is non-zero. Thus, the gradient flow of f pushes γ into the extremal subset $F_1 \cap \ldots \cap F_k$. We assume that there is some $T \in (0, \infty)$ such that $\delta := \Phi_f^T(\gamma) \subseteq F_1 \cap \ldots \cap F_k$. If α_p, α_q denote the gradient curves of f starting at p and q, respectively, the curve δ connects the points $\alpha_p(T)$ and $\alpha_q(T)$. Hence $F_1 \cap \ldots \cap F_k$ is connected.

If such finite T does not exist, perform the construction from above in the space $\Sigma' = d_{F_1}^{-1}([s, a_1])$ for $s > 0$ small enough and $a_1 = \max d_{F_1}$. It is clear that in the space Σ' an appropriate T exists. Now let $s \searrow 0$ and apply Lemma 2.20 on page 39, which gives that $F_1 \cap \ldots \cap F_k$ is connected as Hausdorff limit of the corresponding connected sets $F_1' \cap \ldots \cap F_k'$. □

2.23 Definition. *Let (X, d) be a metric space and $A \subseteq B \subseteq X$. The topological boundary*

$\mathrm{Bd}_B A$ of A in B is defined as follows:

$$\mathrm{Bd}_B A := \{x \in B \mid B_r(x) \cap A \neq \emptyset \text{ and } B_r(x) \cap (B \setminus A) \neq \emptyset \quad \forall r > 0\}$$

where the balls are taken in X.

2.24 Lemma. *Let $M \in \mathrm{ALEX}^n(0)$ be compact and let $F_1, \ldots, F_k$ be a subcollection of some stratification of ∂M with $k < n$. Let $A \subseteq M$ be a convex subset satisfying $A \subseteq F_1 \cap \ldots \cap F_k$ and $\dim A = n - k$. Then $p \in \mathrm{Bd}_{F_1 \cap \ldots \cap F_k} A$ implies that $p \in \partial A$.*

An analog statement holds if $A \subseteq M$ is a convex subset of full dimension. A proof will be carried out in the upcoming book on Alexandrov geometry by Alexander, Kapovitch and Petrunin. The proof of our statement is an adapted version of their one.

Proof. We use induction on $m := n - k$. The base step $m = 1$ is trivial.

For $m \geq 2$ let $p \in \mathrm{Bd}_{F_1 \cap \ldots \cap F_k} A$ and let $\varepsilon > 0$. We endow the extremal set $F_1 \cap \ldots \cap F_k$ with the induced intrinsic metric, denoted by $\hat{d}$. Recall that d and $\hat{d}$ are locally bi-Lipschitz equivalent according to [PP 94, Corollaries 3.2]. We choose $q \in (F_1 \cap \ldots \cap F_k) \setminus A$ such that $\hat{d}(p, q) \leq \frac{\varepsilon}{2}$. Let $x \in A$ be a point with minimal distance $\hat{d}(x, q)$. Let γ be a shortest path in $F_1 \cap \ldots \cap F_k$ from x to q. The choice of x implies that $\gamma^+(0) \notin \Sigma_x A$. In other words, the convex subset $\Sigma_x A \subseteq \Sigma_x$ satisfies $\Sigma_x A \subsetneq \Sigma_x(F_1 \cap \ldots \cap F_k)$. By Lemma 2.22 the set $\Sigma_x(F_1 \cap \ldots \cap F_k) = \Sigma_x F_1 \cap \ldots \cap \Sigma_x F_k$ is connected and therefore $\mathrm{Bd}_{\Sigma_x F_1 \cap \ldots \cap \Sigma_x F_k} \Sigma_x A \neq \emptyset$. By the induction assumption we conclude that $\partial \Sigma_x A \neq \emptyset$ and hence $x \in \partial A$. Since $d(x, p) \leq \hat{d}(x, p) \leq \varepsilon$ and ∂A is closed, this implies that $p \in \partial A$ as desired. $\square$

CHAPTER 3

The Splitting Theorem

3.1 Formulation

This chapter is devoted to the following Splitting Theorem—the main theorem of this work—and its proof.

3.1 Theorem (Splitting Theorem). *Let $M \in \text{Alex}^n(0)$ be compact and let $F_1, \ldots, F_{k+1}$ be a stratification of the boundary ∂M with $k \geq 1$. Assume that the intersection of all boundary strata is empty, i. e.*
$$F_1 \cap \ldots \cap F_{k+1} = \emptyset,$$

and that all intersections of k boundary strata are nonempty, i. e.
$$F_1 \cap \ldots \cap \widehat{F_i} \cap \ldots \cap F_{k+1} \neq \emptyset \quad \forall i \in \{1, \ldots, k+1\}.$$

Then there are Alexandrov spaces $S \in \text{Alex}^{n-k}(0)$ and $D \in \text{Alex}^k(0)$ such that M is isometric to the metric product, i. e. $M \cong S \times D$. Moreover, S has no boundary and satisfies $S \cong F_1 \cap \ldots \cap \widehat{F_i} \cap \ldots \cap F_{k+1}$ $\forall i \in \{1, \ldots, k+1\}$.

3.2 Notation. Throughout the entire chapter we will use the following assumptions and notation.

M and $F_1, \ldots, F_{k+1}$ are given as in the Splitting Theorem.

In addition, for $i = 1, \ldots, k+1$ we define the following:

$$f_i := d_{F_i}, \quad a_i := \max_{p \in M} f_i(p), \quad A_i := \{p \in M \mid f_i(p) = a_i\}$$

3.2 Outline on the proof

The basic steps in the proof of the Splitting Theorem are as follows.

Each intersection of k boundary strata has to be a convex subset. We consider $F_1 \cap \ldots \cap F_k$, say, and obtain by Lemma 2.10 on page 32 that $F_1 \cap \ldots \cap F_k \subseteq A_{k+1}$. If equality holds, we are done, since A_{k+1} is a convex subset. However, in general equality does not hold as the following simple example shows.

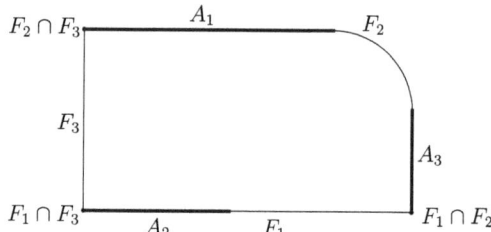

In exchange, it turns out (Theorem 3.4 on page 47) that the Alexandrov space A_{k+1} fulfills the assumptions of the Splitting Theorem with $m+1$ boundary strata, where $m < k$ holds. Moreover, the set $F_1 \cap \ldots \cap F_k$ coincides with the intersection of m boundary strata of A_{k+1}. Thus, we can assume by an induction argument that A_{k+1} has the desired properties and obtain that $F_1 \cap \ldots \cap F_k$ is a convex set. It has no boundary according to Proposition 2.21 on page 40 and will be called a *soul* of M.

In order to obtain a fibration of M into souls, we perform the above argument for each non-collapsed superlevel set M'. This is possible, because such superlevel sets again satisfy the assumptions of the Splitting Theorem by Corollary 2.15 on page 36. The fibration of M into souls will be proved in Proposition 3.6 on page 49. Now we want these souls to be isometric. In order to prove this fact, the gradient flows of the functions f_i are used to push each soul around in M. Again, if some set A_i does not only consist of the appropriate intersection of boundary strata, the flow may not reach the latter set. In this case, we can assume by the induction argument that inside A_i the soul can be pushed to the intersection. Pushing souls can be executed in each superlevel set M', too. Hence, it can be proved that pushing around some soul induces a map homotopic to the identity on the soul. Since the gradient flow is 1-Lipschitz (see Lemma 1.23 on page 17) and the soul has no boundary, pushing the soul is an isometric action. More precisely, Theorem 3.12 on page 53 will provide submetries $\Psi_j \colon M \to F_1 \cap \ldots \cap \widehat{F_j} \cap \ldots \cap F_{k+1}$ for $j = 1, \ldots k+1$, and the restrictions $\Psi_j\big|_S$ to any soul S are isometries.

For each $p \in M$ and $j \in \{1, \ldots, k+1\}$ we will consider the fibers $\Psi_j^{-1}(\Psi_j(p))$, which form a priori different subsets for different indices j. However, it will be shown in Theorem 3.22 on page 60 that they coincide for all indices j. The fibration of M made up by these fibers will turn out in the end to be dual to the fibration into souls. For that reason, the fibers of Ψ_j will be called *dual fibers*. Lytchak's results on submetries in [Lyt 02] will provide many important properties of the dual fibers.

To conclude the proof of the Splitting Theorem, it remains to prove that the souls are equidistant sets. By the results so far, Proposition 3.24 on page 61 will supply this fact along shortest paths to the boundary strata. This means that for each $p \in M$ and $i = 1, \ldots, k+1$ it is known that a shortest path γ from p to F_i has its counterpart in any other dual fiber. In other words, the canonical projection along souls transports γ isometrically into an arbitrary dual fiber. These projections turn out (Proposition 3.18 on page 57) to be locally Lipschitz near almost all points. Therefore, the differential exists and is linear almost everywhere. Vectors tangential to paths γ as above are mapped isometrically by the differential. Moreover, such vectors may span the tangent space of the respective dual fiber. If so, by linearity, the differential is a 1-Lipschitz map. In case the vectors do not span the whole tangent space, it turns out that the remaining subspace comes from some set A'_i (after passing to an appropriate superlevel set M'). Thus, in all directions belonging to A'_i, everything is proved by induction assumption, and also in this case the differential is a 1-Lipschitz map. This will be proved in Proposition 3.28 on page 63. Thereafter it is easy to prove that the projections along souls onto dual fibers are 1-Lipschitz (Theorem 3.29 on page 65). This implies that all souls are equidistant and all dual fibers are isometric convex subsets. The product structure of M follows.

As we have seen, the proof uses induction on k, the number of intersecting boundary strata. In the subsequent section the base step $k = 1$ will be proved.

3.3 First case: two strata

As mentioned in the Introduction, this case is not difficult or surprising. However, we give a proof here for completeness and as a first application of our machinery developed in Chapter 2.

3.3 Theorem. *The Splitting Theorem 3.1 holds for $k = 1$. In this case, the space D is isometric to an interval: $D \cong [0, d]$ where $d = |F_1 F_2|$.*

Proof. Since $F_1 \cap F_2 = \emptyset$ by assumption, Lemma 2.10 on page 32 implies that $F_1 \subseteq A_2$ and $F_2 \subseteq A_1$. Furthermore, equality holds according to Lemma 2.16 on page 36. Thus, we have on the one hand that F_1 and F_2 are convex sets and therefore $F_1, F_2 \in \text{ALEX}^{n-1}(0)$. On the other hand, $|F_1 F_2| = a_1 = a_2 = d_H(F_1, F_2)$ (where d_H denotes the Hausdorff distance), i.e. F_1 and F_2 are equidistant. Proposition 2.21 on page 40 implies that $\partial F_1 = \emptyset$ and $\partial F_2 = \emptyset$.

Let $p \in M \setminus F_1$. According to Proposition 2.14 on page 35, all facts stated above also hold in the space $M' := f_2^{-1}([f_2(p), a_2])$. Hence, the set $S(p) := f_2^{-1}(f_2(p))$ is a convex set of dimension $n-1$ without boundary. We call $S(p)$ the *soul* at p. In addition, for $p \in F_1$ we set $S(p) = F_1$. Thus, the souls form an equidistant fibration of M.

We define the projection $\pi_{F_1} \colon M \to F_1$ in the following way: If $x \in F_1$, then set $\pi_{F_1}(x) = x$. Otherwise let $\pi_{F_1}(x)$ be the endpoint of some shortest path from x to F_1. This shortest path is unique by the equidistance of the souls. Indeed, the path (with inverted direction) is also a shortest path from $\pi_{F_1}(x)$ to $S(x)$, the latter being a boundary stratum of the appropriate space M'. By Lemma 2.11 on page 33 the point $\pi_{F_1}(x)$ is unique, hence the projection π_{F_1} is well-defined. Its restriction to any soul is an isometry, i.e. we claim the following: Let $p \in M$. Then $\pi_{F_1}\big|_{S(p)} \colon S(p) \to F_1$ is an isometry. For $p \in F_1$ the claim is trivial, so let $p \in M \setminus F_1$ and take two distinct points $q, r \in S(p)$. We set $q_1 := \pi_{F_1}(q)$, $r_1 := \pi_{F_1}(r)$. Again by Lemma 2.11 (or alternatively, by the equidistance of the souls), the paths qq_1 and rr_1 intersect each soul perpendicularly. Since all souls are convex sets, it follows that the paths qq_1 and rr_1 are equidistant. In particular, we have that $|q_1 r_1| = |qr|$. Therefore, $S(p)$ embeds isometrically into F_1, and since $S(p)$ is a boundary stratum of some space M', the roles of $S(p)$ and F_1 can be swapped. Hence, the claim is proved.

Now fix an arbitrary shortest path from F_1 to F_2 and denote it by D. Let $\pi_D \colon M \to D$ be the canonical projection along souls. It is clear that each path intersecting the souls perpendicularly is mapped isometrically by π_D. The map $\Psi \colon M \to F_1 \times D$, $x \mapsto (\pi_{F_1}(x), \pi_D(x))$ is bijective. To show that it is an isometry, let $p, q \in M$ be distinct points and let $\gamma \colon I \to M$ be some shortest path from p to q. We set $\gamma_{F_1} := \pi_{F_1}(\gamma)$ and $\gamma_D := \pi_D(\gamma)$ and denote the endpoint of the shortest path from p to $S(q)$ by $r \in S(q)$. Since $\angle prq = \frac{\pi}{2}$, we obtain by Toponogov's Comparison Theorem that

$$|pq|^2 \leq |pr|^2 + |rq|^2 = |\pi_D(p)\, \pi_D(q)|^2 + |\pi_{F_1}(p)\, \pi_{F_1}(q)|^2.$$

On the other hand, for each $t \in I$ and all $h \in \mathbb{R}$ satisfying $t+h \in I$ we have that

$$|\gamma(t)\,\gamma(t+h)|^2 = |\gamma_D(t)\,\gamma_D(t+h)|^2 + |\gamma_{F_1}(t)\,\gamma_{F_1}(t+h)|^2 + o(h),$$

which implies by integration that

$$L(\gamma)^2 = L(\gamma_D)^2 + L(\gamma_{F_1})^2.$$

3.4. Fibration into souls

Thus, we obtain that

$$L(\gamma)^2 = |pq|^2 \leq |\pi_D(p)\,\pi_D(q)|^2 + |\pi_{F_1}(p)\,\pi_{F_1}(q)|^2 \leq L(\gamma_D)^2 + L(\gamma_{F_1})^2 = L(\gamma)^2$$

and therefore everywhere equality. This implies that Ψ is an isometry. □

The proof of the Splitting Theorem for $k \geq 2$ is carried out in the rest of this chapter.

3.4   Fibration into souls

In this section we prove that the space M is fibrated into convex subsets of codimension k without boundary. The key result, which is essential in further sections, too, is the following.

3.4 Theorem. *Let $\ell \in \{1, \ldots, k+1\}$. The boundary ∂A_ℓ fulfills the same assumptions as ∂M, i.e. there is a stratification $G_1, \ldots, G_{m+1}$ of ∂A_ℓ with $m < k$ such that $G_1 \cap \ldots \cap G_{m+1} = \emptyset$ and $G_1 \cap \ldots \cap \widehat{G_j} \cap \ldots \cap G_{m+1} \neq \emptyset \quad \forall j \in \{1, \ldots, m+1\}$.
More precisely, $\partial A_\ell \neq \emptyset$ holds if and only if $A_\ell \not\supseteq F_1 \cap \ldots \cap \widehat{F_\ell} \cap \ldots \cap F_{k+1}$. In this case let $I \subseteq \{1, \ldots, k+1\}$ be the maximal set of indices such that $A_\ell \subseteq \bigcap_{i \in I} F_i =: F$. Then $1 \leq |I| < k$ and the strata of ∂A_ℓ are given by $A_\ell \cap F_i$ for $i \in \{1, \ldots, k+1\} \setminus (I \cup \{\ell\})$ plus the stratum $\mathrm{Bd}_F A_\ell$. In addition, the dimension of A_ℓ satisfies $\dim A_\ell = \dim F = n - |I|$.*

Proof. The assertion will be proved for A_{k+1}. By Lemma 2.10 on page 32 we have that $F_1 \cap \ldots \cap F_k \subseteq A_{k+1}$. If equality holds, it follows from Proposition 2.21 on page 40 that $\partial A_{k+1} = \emptyset$. Hence, we assume that $F_1 \cap \ldots \cap F_k \subsetneq A_{k+1}$. Let $p \in A_{k+1}$ be a point such that $f_i(p) \neq 0$ for a maximal number of indices $i \in \{1, \ldots, k\}$. After renumbering the boundary strata we may assume that $f_i(p) \neq 0$ for $i = 1, \ldots, m$ and $f_i(p) = 0$ for $i = m+1, \ldots, k$. According to Lemma 2.16 on page 36, we have that $m < k$.

We claim that $f_{m+1}(q) = \ldots = f_k(q) = 0 \quad \forall q \in A_{k+1}$. Assume, by way of contradiction, there exists $q \in A_{k+1}$ and $j \in \{m+1, \ldots, k\}$ with $f_j(q) \neq 0$. Note that $f_i(x) = 0$ is equivalent to $x \in F_i$ and F_i is an extremal set. Thus, any inner point r of some shortest path pq fulfills $f_i(r) \neq 0$ for $i \in \{1, \ldots, m\} \cup \{j\}$. Since A_{k+1} is a convex set, we obtain that $r \in A_{k+1}$, contradicting the choice of p.

By the claim we have that $A_{k+1} \subseteq F_{m+1} \cap \ldots \cap F_k$ and therefore $\dim A_{k+1} \leq \dim(F_{m+1} \cap \ldots \cap F_k) = n - (k-m)$, see Theorem 2.18 on page 37. On the other hand, if we set $s_i := f_i(p)$ for $i = 1, \ldots m$, again Lemma 2.16 implies that $f_1^{-1}([0, s_1)) \cap \ldots \cap f_m^{-1}([0, s_m)) \cap F_{m+1} \cap \ldots \cap F_k \subseteq A_{k+1}$. We conclude that $n - (k-m) \leq \dim A_{k+1}$. This gives that

$$\dim A_{k+1} = \dim(F_{m+1} \cap \ldots \cap F_k) = n - (k-m).$$

47

Now we set $G_i := A_{k+1} \cap F_i$ for $i = 1, \ldots, m$. Then we have that $f_i\big|_{A_{k+1}} = d_{G_i}$ for $i = 1, \ldots, m$. This fact can be shown like in the proof of Lemma 2.13 on page 34. We want to show that, say, G_1 is a boundary stratum of A_{k+1}. First, G_1 is extremal in A_{k+1}, because it is closed and for each $p \in A_{k+1} \setminus G_1$ the following holds: If the function $d_p\big|_{F_1}$ attains its minimum at $q \in F_1$, then in fact $q \in G_1$, and hence $\angle pqx \leq \frac{\pi}{2}$ $\forall x \in A_{k+1}$.

It remains to prove that G_1 has locally constant dimension $n - (k - m) - 1$; then G_1 is a stratum of ∂A_{k+1}. Let $t > 0$ be small enough such that it is a regular value of the function d_{G_1}. By the Morse Lemma, the level set $Z := d_{G_1}^{-1}(t)$ is of locally constant dimension and has codimension 1 in A_{k+1}. We define a map $\psi \colon Z \to F_1$ like follows: For $z \in Z$ let $\psi(z)$ be the endpoint of some shortest path from z to F_1. We claim that ψ is noncontracting. Indeed, by Lemma 2.11 on page 33 each shortest path from some point in Z to F_1 is perpendicular to F_1, and the same holds for the boundary stratum F_1' in any superlevel set $M' = f_1^{-1}([t', a_1])$, $t' \in (0, t)$. Hence, the distance of two corresponding points on two shortest paths from $z_1, z_2 \in Z$ to F_1 cannot decrease. Therefore, $\dim \psi(Z) \geq n - (k - m) - 1$ and since $\psi(Z) \subseteq G_1 \subseteq F_1 \cap F_{m+1} \cap \ldots \cap F_k$, the extremal set G_1 has locally constant dimension $n - (k - m) - 1$.

To complete the proof we have to show that there exists the additional boundary stratum $G_{m+1} := \text{Bd}_{F_{m+1} \cap \ldots \cap F_k} A_{k+1}$ and that the strata intersect like stated in the Theorem.

Let $p \in \partial A_{k+1}$. Assume that $p \notin G_{m+1}$, which implies that $T_p A_{k+1} = T_p(F_{m+1} \cap \ldots \cap F_k)$. According to Proposition 2.21 on page 40, we have that $p \in G_i$ for some $i \in \{1, \ldots, m\}$. Indeed, this is true if $T_p A_{k+1} \approx \mathbb{R}^{n-(k-m)-1} \times \mathbb{R}_+$ and such points are dense in $\partial A_{k+1} \setminus G_{m+1}$. Therefore it follows that $\partial A_{k+1} \subseteq G_1 \cup \ldots \cup G_{m+1}$. It is clear that $G_{m+1} \neq \emptyset$, because each limit point of some gradient curve to the function f_{k+1} starting (and hence staying) in $F_{m+1} \cap \ldots \cap F_k$ lies in G_{m+1}. In order to refine this argument, the same holds if the gradient curve starts in $F_1 \cap \ldots \cap \widehat{F_j} \cap \ldots \cap F_k$, for each $j \in \{1, \ldots, m\}$. By slightly shifting the strata $F_1, \ldots, F_m$, i.e. performing that construction in superlevel sets M' close to M, we conclude that there are points in G_{m+1} arbitrary close to $G_1 \cap \ldots \cap \widehat{G_j} \cap \ldots \cap G_m$, but not contained in any $G_1, \ldots, G_m$, for each $j \in \{1, \ldots, m\}$. According to Lemma 2.24 on page 42, we have that $G_{m+1} \subseteq \partial A_{k+1}$. Hence, it follows that $\overline{\partial A_{k+1} \setminus (G_1 \cup \ldots \cup G_m)}$ is nonempty and therefore a boundary stratum. Moreover, we have that $\overline{\partial A_{k+1} \setminus (G_1 \cup \ldots \cup G_m)} \subseteq G_{m+1}$.

In order to see equality, consider the doubling $\bar{M}$ by gluing along $F_1 \cup \ldots \cup F_m$. The set $\bar{A}_{k+1}$ is a convex set of $\bar{M}$. Its boundary $\partial \bar{A}_{k+1}$ coincides on the one hand with the topological boundary $\text{Bd}_{\bar{F}_{m+1} \cap \ldots \cap \bar{F}_k} \bar{A}_{k+1} = \bar{G}_{m+1}$ and on the other hand with the doubling of the stratum $\overline{\partial A_{k+1} \setminus (G_1 \cup \ldots \cup G_m)}$.

Similar doubling procedures show that $G_1 \cup \ldots \cup G_{m+1}$ is indeed a stratification of ∂A_{k+1}, i.e. there is no intersection $G_i \cap G_j$, $i \neq j$ which is itself a boundary stratum.

3.4. Fibration into souls

As proved above, we have that

$$G_1 \cap \ldots \cap \widehat{G_j} \cap \ldots \cap G_m \cap G_{m+1} \neq \emptyset \quad \forall j \in \{1, \ldots, m\}.$$

Moreover, by $G_1 \cap \ldots \cap G_m = F_1 \cap \ldots \cap F_m \cap A_{k+1} \subseteq F_1 \cap \ldots \cap F_k \subseteq F_1 \cap \ldots \cap F_m \cap A_{k+1}$ we conclude that

$$G_1 \cap \ldots \cap G_m = F_1 \cap \ldots \cap F_k \neq \emptyset.$$

It remains to show that $G_1 \cap \ldots \cap G_{m+1} = \emptyset$. Again, choose $p \in A_{k+1}$ such that $s_i := f_i(p) > 0 \quad \forall i \in \{1, \ldots, m\}$. Then $f_1^{-1}([0, s_1)) \cap \ldots \cap f_m^{-1}([0, s_m)) \cap F_{m+1} \cap \ldots \cap F_k \subseteq A_{k+1}$ is a neighborhood of the set $G_1 \cap \ldots \cap G_m$. In particular, this neighborhood does not contain any points of G_{m+1}. This proves the assertion. □

3.5 Corollary. *Each intersection $F_1 \cap \ldots \cap \widehat{F_j} \cap \ldots \cap F_{k+1}$, $j \in \{1, \ldots, k+1\}$ is a convex subset without boundary. Moreover, each point $p \in M$ lies in a convex subset $S \subseteq M$ of dimension $n - k$ without boundary.*

Proof. Let $j \in \{1, \ldots, k+1\}$. If $F_1 \cap \ldots \cap \widehat{F_j} \cap \ldots \cap F_{k+1} = A_j$, the first statement is proved. Otherwise, according to Theorem 3.4, the intersection $F_1 \cap \ldots \cap \widehat{F_j} \cap \ldots \cap F_{k+1}$ coincides with the intersection of m corresponding boundary strata of A_j with $m < k$ and $\dim A_j = n - (k - m) = n - k + m$. Hence, induction on the number of intersecting boundary strata proves the statement.

In order to prove the second statement, let $p \in M$ be an arbitrary point and set $s_i := f_i(p)$ for $i = 1, \ldots, k+1$. If there is an index j such that $s_j = a_j$, then $p \in A_j$, and the statement follows like above by induction assumption. If such an index does not exist, p is not contained in any intersection of k boundary strata. Hence, there is an index (in fact, at least two indices) j such that $0 < s_j < a_j$. Then consider the superlevel set $M' := f_j^{-1}([s_j, a_j])$ and iterate the procedure, which clearly terminates. □

Note that the Corollary does not yet guarantee a fibration of M, because the procedure described in the proof may depend on the order of passing to superlevel sets. Indeed, problems may occur exactly if $p \in M$ lies in two different sets A_i, A_j. By the induction assumption, both sets posses unique fibrations, but the respective fibers containing p could form different subsets of M. Therefore, we have to examine the intersection $A_i \cap A_j$ and to prove that the fibrations of A_i and A_j in fact match.

3.6 Proposition. *The space M is fibrated into convex subsets of dimension $n - k$ without boundary, called souls. On each soul the functions $f_1, \ldots, f_{k+1}$ are constant. In particular, all intersections of k boundary strata are souls. Moreover, each set A_i, $i \in \{1, \ldots, k+1\}$ is* fibrated into souls as well, and all these souls are also souls of M. The same holds for each corresponding set A'_i in any non-collapsed superlevel set $M' \subseteq M$.

Proof. The fibration is obtained as described in the proof of Corollary 3.5. Note that according to Theorem 3.4 and its proof, in each set A_i all but one boundary strata are obtained by cutting down the strata of ∂M to A_i. Moreover, the distance function to such stratum of ∂A_i is nothing but the corresponding function f_j restricted to A_i. This fact ensures that all functions f_j are constant on each soul of each set A_i. This in turn carries over to each A'_i in any superlevel set $M' \subseteq M$.

As mentioned above, it remains to prove the following claim: Assume that $p \in M$ is contained in several sets A_i. Then the fibration of each A_i gives a soul for p, but all these souls coincide.

In order to prove the claim, assume without loss of generality that $p \in A_1 \cap A_2$. First we consider A_1. According to Theorem 3.4, the set $A_1 \cap F_2$ is a stratum of ∂A_1 and we have that $d_{A_1 \cap F_2}(p) = f_2(p) = a_2$. On the other hand, if $q \in A_1$ is some point at maximal distance to $A_1 \cap F_2$, it follows that $a_2 \geq d_{A_1 \cap F_2}(q) \geq d_{A_1 \cap F_2}(p) = a_2$. This implies that $q \in A_1 \cap A_2$. Therefore, the set $A := A_1 \cap A_2$ coincides with the set of points in A_1 at maximal distance to the boundary stratum $A_1 \cap F_2$.

Analogously, A coincides with the set of points in A_2 at maximal distance to $A_2 \cap F_1$. By applying Theorem 3.4 to the set A inside A_1 or inside A_2, respectively, we can determine the stratification of ∂A in two ways. They do not have to coincide, but the stratifications into *primitive* boundary strata coincide, since they are uniquely determined by the boundary ∂A itself. By induction, we may assume that the Splitting Theorem is proved for the space A. We apply Corollary 4.2 on page 67 and obtain that there are souls of A which coincide as subsets whether $A \subseteq A_1$ is considered or $A \subseteq A_2$. Indeed, if $S \subseteq A$ is given as the intersection of boundary strata coming from A_1, then in fact S is already given as the intersection of primitive boundary components. This follows for dimensional reasons by Corollary 4.1 on page 67. Hence, the same is also true for the stratification coming from A_2.

Now the claim follows, because all other souls of A can be recovered from the soul S. Recall that the product structure of A is known by induction assumption. Therefore, the dual fiber through some regular $p \in S$ is given as the preimage of p under orthogonal projection onto S. Furthermore, two points $q_1, q_2 \in A$ lie in the same soul if and only if $|q_1 q_2|$ coincides with the distance between the respective dual fibers. Since almost all dual fibers are recovered, all souls can be recovered by taking closures. Thus, the fibration of A is independent whether $A \subseteq A_1$ or $A \subseteq A_2$ is considered. □

3.7 Remark. It is possible that the Splitting Theorem can be proved without unique souls in the sense from above. Since by induction assumption each A_i possesses a fibration into souls,

3.5. Isometry of the souls via gradient flow

on intersections $A_i \cap A_j$ one could simply decide for one. But then in the subsequent results and proofs one always has to take into account in which set A_i some soul is obtained. This causes at least some trouble and is the reason why we decided to prove the uniqueness of the fibration first.

The fibration of M into souls enables the following notation.

3.8 Notation. For each $p \in M$ the souls of M containing p is denoted by $S(p)$.

In the following section we prove the fact that all souls of M are isometric and examine the respective isometries.

3.5 Isometry of the souls via gradient flow

First we will establish that certain souls are isometric, namely all intersections of k boundary strata. Passing to superlevel sets then implies that in fact all souls are isometric. However, the obtained isometries are too abstract for later use. We need the following fact: For each soul S which coincides with an intersection of k boundary strata (there are $k+1$ such souls) there exists a submetry (see Definition 3.11 below) $\Psi: M \to S$ which maps each soul of M isometrically onto S. The canonical candidate for Ψ is the gradient flow, of course. More precisely, if $S = A_1$, say, then Ψ should be given by $\Phi_{f_1}^T$ for T big enough. If such finite T does not exist, some limit is taken. Moreover, if $S \subsetneq A_1$, the gradient flow has to be composed with the flow inside A_1, whose existence can be assumed by induction. The details will be given later. We start with the following

3.9 Proposition. *All souls of the form $S_j := F_1 \cap \ldots \cap \widehat{F_j} \cap \ldots \cap F_{k+1}$ with $j \in \{1, \ldots, k+1\}$ are isometric.*

Proof. The assertion will be proved for the souls S_1 and S_{k+1}. We start with the following special case.

Assume, there is $T \in (0, \infty)$ such that $\Phi_{f_{k+1}}^T(S_1) \subseteq S_{k+1}$ and $\Phi_{f_1}^T(S_{k+1}) \subseteq S_1$. Then we have that $\Phi := \Phi_{f_1}^T \circ \Phi_{f_{k+1}}^T : S_1 \to S_1$ is nonexpanding (by Lemma 1.23 on page 17). We claim that Φ is homotopic to the identity on S_1. For the proof we apply the gradient flows in superlevel sets obtained by shifting the boundary stratum F_1. More precisely, for $t \in [0, 1)$ let $M_t := f_1^{-1}([ta_1, a_1])$ and define Φ_t analogously to Φ from above, but in the space M_t. This is possible, because the gradient curves to f_1 and f_{k+1} all start and hence stay in $F_2 \cap \ldots \cap F_k$, recall Lemma 2.9 on page 32. Moreover, if in the space M the pushed soul lies in the other soul after time T, the same holds in all superlevel sets M_t. It is clear that $\Phi_0 = \Phi$ and in addition we set $\Phi_1 := \mathrm{id}_{S_1}$. In order to show that $(t, p) \mapsto \Phi_t(p)$ is a homotopy, it is sufficient to show that $t \mapsto \Phi_t(p)$ is continuous for each fixed $p \in S_1$. Indeed, since all Φ_t are nonexpanding, if $t \in [0,1]$, $p \in S_1$ and $\rho > 0$ are given, we can then choose $\delta > 0$ such that $\Phi_{B_\delta(t)}(p) \subseteq B_{\rho/2}(\Phi_t(p))$. Then for all $t' \in B_\delta(t)$, $p' \in B_{\rho/2}(p)$ we have that

$$|\Phi_{t'}(p')\,\Phi_t(p)| \leq |\Phi_{t'}(p')\,\Phi_{t'}(p)| + |\Phi_{t'}(p)\,\Phi_t(p)| \leq \frac{\rho}{2} + \frac{\rho}{2} = \rho.$$

In order to see that $t \mapsto \Phi_t(p)$ is continuous for any fixed $p \in S_1$, note that $s \mapsto \Phi_{f_{k+1}}^s$ is continuous and that f_{k+1} stays the same in all superlevel sets M_t. In addition, Φ_{f_1} is nonexpanding and f_1 changes only by an additive constant in the superlevel sets M_t. Hence the claim is proved.

We show now that Φ is an isometry. By Proposition 3.6 on page 49, the soul S_1 is an Alexandrov space of dimension $n - k$ without boundary. According to [GP 93, Lemma 1], the $(n-k)$th Alexander-Spanier cohomology of S_1 is non-trivial, i.e. $\bar{H}^{n-k}(S_1, \mathbb{Z}_2) = \mathbb{Z}_2$. On the other hand, for each $p \in S_1$ we have that $\bar{H}^{n-k}(S_1 \setminus \{p\}, \mathbb{Z}_2) = 0$. Indeed, using the gradient flow of the function d_p^2, we get that $S_1 \setminus \{p\}$ is homotopically equivalent to some space of dimension less than $n - k$. Therefore, the map Φ is surjective, because the homotopy $\Phi \simeq \mathrm{id}_{S_1}$ implies that the induced homomorphism $\bar{H}^{n-k}(\Phi) \colon \mathbb{Z}_2 \to \mathbb{Z}_2$ is an isomorphism. Being a surjective nonexpanding map of a compact metric space onto itself, Φ is an isometry (see e.g. [BBI 01, Theorem 1.6.15]).

It remains to deal with the case that such finite time T as above does not exist. We will establish later (or by the Splitting Theorem) that all souls of M are isometric. Hence, we can assume by induction that this fact is already proved for the sets A_1 and A_{k+1}. If $A_1 \cap A_{k+1} \neq \emptyset$, take some point $p \in A_1 \cap A_{k+1}$ and consider the soul $S(p)$. According to Proposition 3.6, the souls S_1 and S_{k+1} are also souls of the spaces A_1 and A_{k+1}, respectively, and $S(p)$ is a soul of both spaces. This implies that $S_1 \cong S(p) \cong S_{k+1}$ as desired.

Thus, we may assume that $A_1 \cap A_{k+1} = \emptyset$. Choose a point $p \in (F_2 \cap \ldots \cap F_k) \setminus (A_1 \cup A_{k+1})$ and consider the superlevel set $M' := f_1^{-1}([f_1(p), a_1])$. Then there exists some $T \in (0, \infty)$ such that $\Phi_{f_{k+1}}^T(S_1) \subseteq S'_{k+1} := F'_1 \cap \ldots \cap F'_k$. Indeed, since S_1 is compact, such T exists for each finite ε-net N of S_1, and since the gradient flow is nonexpanding, all other points are pushed ε-close to $\Phi_{f_{k+1}}^T(N)$. It is clear by construction that $A'_1 = A_1$ and $A'_1 \cap A'_{k+1} = \emptyset$, where A'_1, A'_{k+1} are defined as A_1, A_{k+1}, but for the space M'. Therefore, we can pass to some superlevel set $M'' \subseteq M'$ by shifting the boundary stratum F'_{k+1} in an analogous way and obtain that the space M'' fulfills the assumptions from the first part of the proof. In particular, the corresponding souls S''_1 and S''_{k+1} of M'' are isometric. This in turn holds if we choose p from above arbitrarily close to A_{k+1}. By Lemma 2.20 on page 39, the particular souls S''_{k+1} converge to a soul of A_{k+1}. Since all souls of A_{k+1} are isometric by induction assumption, we have that $S''_{k+1} \cong S_{k+1}$. The

3.5. Isometry of the souls via gradient flow

same construction applied to S_1'' gives that $S_1 \cong S_1''$ and therefore $S_1 \cong S_{k+1}$. $\square$

For completeness we mention here the following consequence.

3.10 Corollary. *All souls of M are isometric.*

Proof. Let $S \subseteq M$ be a soul. It is sufficient to prove that $S \cong S_1 := F_2 \cap \ldots \cap F_{k+1}$. If $S \subseteq A_1$, the assertion follows by induction assumption. Therefore we may assume that $s_1 := f_1(S) < a_1$ and consider the superlevel set $M' := f_1^{-1}([s_1, a_1])$. Proposition 3.9 implies that $S_1 \cong F_1' \cap F_3' \cap \ldots \cap F_{k+1}'$. If $S \subseteq A_2'$, the assertion is proved; otherwise iterate the process of passing to superlevel sets. $\square$

As announced before, the constructed isometries are not sufficient for our purpose. We need more information about the action of the gradient flows. First we recall the definition of submetries.

3.11 Definition. Let X and Y be metric spaces. A map $f\colon X \to Y$ is a *submetry* if $f(\bar{B}_r(x)) = \bar{B}_r(f(x))$ holds for all $x \in X$ and all $r \geq 0$.

This definition traces back to Berestovskii and generalizes Riemannian submersions to metric spaces. Lytchak obtained in [Lyt 02] important results about submetries of Alexandrov spaces. Some of them will be used later. First we formulate the main result of the current section, which is crucial for the proof of the Splitting Theorem.

3.12 Theorem. *For each soul of the form $S_j := F_1 \cap \ldots \cap \widehat{F_j} \cap \ldots \cap F_{k+1}$, $j \in \{1, \ldots, k+1\}$ there exists a submetry $\Psi_j\colon M \to S_j$ such that the following holds: If $S \subseteq M$ is a soul, the restricted map $\Psi_j\big|_S\colon S \to S_j$ is an isometry.*
Moreover, Ψ_j is composed of gradient flows like follows: $\Phi_{f_j}^T$ is a submetry of M onto A_j, possibly with $T = \infty$. The latter case is defined via (possibly not uniquely determined) limits. If $S_j \subsetneq A_j$, the flowing process is iterated inside A_j and hence, Ψ_j is obtained as the composition of these gradient flows. In addition, if $S \subseteq M$ is a soul and $t \in [0, \infty]$, the set $\Phi_{f_j}^t(S)$ is also a soul of M.

The proof of the Theorem uses the subsequent lemmata. They provide basic results and concepts for the proof.

3.13 Lemma. *Theorem 3.12 holds for $k = 1$.*

Proof. This follows immediately from the proof of Theorem 3.3 on page 45. In particular, T can be chosen as $T = d$. Indeed, let $p \in M \setminus \partial M$ and take shortest paths from p to F_1 and F_2. Denote the endpoints of these paths by q_1 and q_2, respectively. Then the equidistance of the souls implies that $p \in q_1 q_2$. Hence, $|\nabla_p(f_i)| = 1$ for $i = 1, 2$. It follows that $\Phi^d_{f_1}(M) = F_2$ and vice versa $\Phi^d_{f_2}(M) = F_1$. Moreover, for each soul $S \subseteq M$ the restriction $\Phi^d_{f_i}\big|_S$ is an isometry for $i = 1, 2$. Since the gradient flows $\Phi^d_{f_i}$ are 1-Lipschitz, this implies that they are submetries. The last assertion of Theorem 3.12 follows by passing to superlevel sets. □

3.14 Lemma. *For $k \geq 2$ Theorem 3.12 holds on each intersection of $k-1$ boundary strata. More precisely, all statements of Theorem 3.12 about Ψ_j hold if all occurring sets are intersected with $F_1 \cap \ldots \cap \widehat{F_i} \cap \ldots \cap \widehat{F_j} \cap \ldots \cap F_{k+1}$ for any $i \neq j$.*

Proof. We prove the Lemma for $j = 1$ and $i = k+1$, i.e. we construct a submetry $\Psi_1 \colon F_2 \cap \ldots \cap F_k \to S_1 := F_2 \cap \ldots \cap F_{k+1}$. By induction assumption, where the base step is provided by Lemma 3.13, we may assume that Theorem 3.12 is proved on A_1. The map Ψ_1 is now constructed as described in this Theorem. If there is no finite T such that $\Phi^T_{f_1}(F_2 \cap \ldots \cap F_k) = S_1$, the gradient flow can be extended to $T = \infty$ like in the proof of Proposition 3.9 on page 51. This proof also implies that $\Psi_1\big|_{S_{k+1}} \colon S_{k+1} \to S_1$ is an isometry. In order to show this fact for an arbitrary soul $S \subseteq F_2 \cap \ldots \cap F_k$, let such soul S be given. If $S \subseteq A_1$, there is nothing to show by induction assumption. Otherwise, passing to the superlevel set $M' := f_1^{-1}([s_1, a_1])$, where $s_1 := f_1(S)$, leads to $S = F'_1 \cap \ldots \cap F'_k$, and again Proposition 3.9 can be applied. Hence, also the restriction $\Psi_1\big|_S \colon S \to S_1$ is an isometry.

It remains to show that for any $t \geq 0$ and any soul $S \subseteq F_2 \cap \ldots \cap F_k$ the set $\Phi^t_{f_1}(S)$ is also a soul. It is sufficient to prove this for $S = S_{k+1}$, because otherwise one can argue like above in superlevel sets or by induction assumption if $S \subseteq A_1$. Note that the following arguments only rely on the properties about Ψ_1 that were just proved. Therefore, they will also apply in the proof of Theorem 3.12 later.

Let $p, q \in S_{k+1}$ be distinct points and let $P := \Psi_1^{-1}(\Psi_1(p))$, $Q := \Psi_1^{-1}(\Psi_1(q))$ denote the fibers of the submetry Ψ_1 passing through p or q, respectively. We claim the following: P and Q are equidistant with $|PQ| = |pq|$, and points $x \in P$, $y \in Q$ fulfilling $|xy| = |PQ|$ have to lie in the same soul $S(x) = S(y)$. In order to prove the claim, let γ be some shortest path from P to Q. Since Ψ_1 is 1-Lipschitz, we obtain that $L(\Psi_1(\gamma)) \leq L(\gamma)$ and therefore equality, because also $\Psi_1(\gamma)$ is a path from P to Q. Moreover, $\Psi_1\big|_{S_{k+1}} \colon S_{k+1} \to S_1$ is an isometry, which implies that $|pq| = L(\Psi_1(\gamma)) = L(\gamma) = |PQ|$. For the equidistance let $x \in P$ and $y := S(x) \cap Q$, the latter being well-defined since $\Psi_1\big|_{S(x)}$ is an isometry. This clearly also implies that $|xy| = |PQ|$. By swapping the roles of x and y, the equidistance of P and Q is proved. Now assume, by way of contradiction, there is $z \in Q$ satisfying $|xz| = |xy|$ and $z \notin S(x) = S(y)$. Then there is some $s \in xz$ such that $s \notin S(x) \cup S(z)$. Let $u := S(s) \cap Q$. This implies, according to the results above, that $|su| \leq |sz|$. Hence, we have that

$$|xz| = |xs| + |sz| \geq |xs| + |su| \geq |xu| \geq |xQ| = |xy| = |xz|$$

3.5. Isometry of the souls via gradient flow

and therefore everywhere equality. But this means that the shortest paths xu and xz branch at the point s, which is a contradiction.

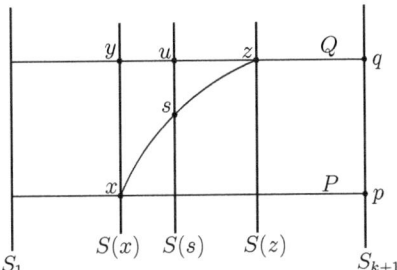

Thus, the claim is proved.

Now let $t \geq 0$. Since $\Psi_1|_{S_{k+1}}$ is an isometry and $\Phi^t_{f_1}$ is 1-Lipschitz, the restriction $\Phi^t_{f_1}|_{S_{k+1}}$ is also an isometry onto its image. In other words, we have that $|\Phi^t_{f_1}(p)\,\Phi^t_{f_1}(q)| = |pq|$, and the claim implies that the points $\Phi^t_{f_1}(p)$ and $\Phi^t_{f_1}(q)$ lie in the same soul, as desired. $\square$

As a consequence, there are continuous paths between any two souls of M in the space of souls. More precisely, we obtain the following result.

3.15 Lemma. *Let S, S' be souls of M. Then there exists a continuous map $\Upsilon \colon S \times [0,1] \to M$ such that $\Upsilon^\tau(S) := \Upsilon(S, \tau)$ is a soul for each $\tau \in [0,1]$ and $\Upsilon^0(S) = S$, $\Upsilon^1(S) = S'$.*

Proof. We combine the proofs of Lemma 3.14 and Corollary 3.10. The latter one shows how passing to non-collapsed superlevel sets reduces the proof to the case that S and S' are given as intersections of k boundary strata. This case, in turn, is proved by Lemma 3.14. Note that the gradient flow $\Phi^t_{f_i}$ on some strata intersection $F_1 \cap \ldots \cap \widehat{F_i} \cap \ldots \cap \widehat{F_j} \cap \ldots \cap F_{k+1}$, $i \neq j$ has a unique extension to $t = \infty$ via limits. This was shown in the proof of Proposition 3.9 on page 51 and relies on the fact that converging souls are in this case given as intersections of $k - 1$ fixed boundary strata and one converging stratum. Thus, according to Lemma 2.20 on page 39, the Hausdorff limits exist. (Or we encounter sets A_i, for which everything is proved by induction assumption.) By reparametrization $[0, \infty] \to [0, 1]$, we obtain the desired map Υ. Hence, by concatenation, such map Υ exists for any pair of souls. $\square$

Now Theorem 3.12 on page 53 can be proved.

Proof of Theorem 3.12. The proof is carried out for $j = 1$, i.e. for $\Psi_1 \colon M \to S_1$. Let $S \subseteq M$ be a soul. First we consider the case that there is some finite $T \geq 0$ such that $\Phi^T_{f_1}(M) = S_1$.

According to Lemma 3.15, there is a continuous map $\Upsilon \colon S_{k+1} \times [0,1] \to M$ such that $\Upsilon^0(S_{k+1}) = S_{k+1}$ and $\Upsilon^1(S_{k+1}) = S$. In addition, on $F_2 \cap \ldots \cap F_k$ the existence of Ψ_{k+1}

Chapter 3. The Splitting Theorem

is proved by Lemma 3.14. Thus, the following map is well-defined for each $\tau \in [0,1]$ and nonexpanding:

$$S_{k+1} \xrightarrow{\Upsilon^\tau} \Upsilon^\tau(S_{k+1}) \xrightarrow{\Phi^T_{f_1}} S_1 \xrightarrow{\Psi_{k+1}} S_{k+1} \quad (*)$$

Therefore, the map for $\tau = 1$ is homotopic to

$$S_{k+1} \xrightarrow{\Phi^T_{f_1}} S_1 \xrightarrow{\Psi_{k+1}} S_{k+1}$$

which pushes the soul S_{k+1} around inside the intersection $F_2 \cap \ldots \cap F_k$. Again by Lemma 3.14, this map is an isometry. The argument from the proof of Proposition 3.9 on page 51 involving the Alexander-Spanier cohomology now implies that all maps in $(*)$ are isometries, in particular $\Phi^T_{f_1}\big|_S : S \to S_1$. The fact that $\Phi^t_{f_1}(S)$ is a soul for each $t \geq 0$ follows like in the proof of Lemma 3.14. It was pointed out there that the argument relies only on the properties of Ψ_1, which are established now on M.

It remains to examine the case that such finite T does not exist. If at least $\Phi^T_{f_1}(M) = A_1$ for some finite T, then the map Ψ_1 is constructed as the composition $\Psi_1^{A_1} \circ \Phi^T_{f_1}$. Here, $\Psi_1^{A_1} : A_1 \to S_1$ denotes the respective submetry inside A_1, which exists by induction assumption. Otherwise, we have to show that the 1-Lipschitz property of $\Phi^t_{f_1}$ carries over to $t = \infty$. Note that our aim is to construct some map $\Phi^\infty_{f_1} : M \to A_1$, but it will not extend the flow $\Phi^t_{f_1}$ continuously in the t-parameter! Moreover, the constructed map will depend on some choices. On intersections of $k-1$ boundary strata, the situation was different, as was pointed out in the proof of Lemma 3.15.

Let $(p_\ell)_{\ell \in \mathbb{N}} \subseteq M$ be a dense sequence of points and $0 \leq t_{0,1} < t_{0,2} < t_{0,3} < \ldots$ a diverging sequence in $\mathbb{R}$. Since M is compact, the sequence $\left(\Phi^{t_{0,i}}_{f_1}(p_1)\right)$ has some partial limit, denoted by $\Phi^\infty_{f_1}(p_1)$. Let $(t_{1,i})$ be a subsequence of $(t_{0,i})$ such that $\Phi^{t_{1,i}}_{f_1}(p_1) \xrightarrow{i \to \infty} \Phi^\infty_{f_1}(p_1)$. Now define $\Phi^\infty_{f_1}(p_2)$ as some partial limit of $\left(\Phi^{t_{1,i}}_{f_1}(p_2)\right)$, and so on. It is clear by continuity that $\Phi^\infty_{f_1}(p_\ell) \in A_1 \ \forall \ell \in \mathbb{N}$. We claim that $\Phi^\infty_{f_1} : \{p_\ell \mid \ell \in \mathbb{N}\} \to A_1$ is nonexpanding. Indeed, assume that, say, $|p_1 p_2| = d$ and $\left|\Phi^\infty_{f_1}(p_1)\, \Phi^\infty_{f_1}(p_2)\right| > d$. Then there exists $t_{2,i}$ for i big enough such that $\left|\Phi^{t_{2,i}}_{f_1}(p_1)\, \Phi^{t_{2,i}}_{f_1}(p_2)\right| > d$, contradiction.

Now, since A_1 is compact, in particular complete, there exists a (unique) nonexpanding extension $\Phi^\infty_{f_1} : M \to A_1$. All arguments of the first part of the proof carry over, if we substitute $\Phi^T_{f_1}$ by $\Psi_1^{A_1} \circ \Phi^\infty_{f_1}$. In particular, $\Phi^\infty_{f_1}\big|_S$ is an isometry onto its image for each soul S. This holds independently of the choice of our dense sequence (p_ℓ). Therefore, we can conclude that also $\Phi^t_{f_1}\big|_S$ is an isometry onto its image for all $t \geq 0$. Indeed, assume there are points $x, y \in S$ whose distance is decreased by $\Phi^t_{f_1}\big|_S$. Then we substitute p_1, p_2 by x, y and obtain for the corresponding construction of $\Phi^\infty_{f_1}$ that $\left|\Phi^\infty_{f_1}(x)\, \Phi^\infty_{f_1}(y)\right| < |xy|$, contradiction. Finally, by these results, also the fact that $\Phi^t_{f_1}(S)$ is a soul for each $t \in [0, \infty]$ and each soul S carries over. $\square$

3.16 Corollary. Let $j \in \{1, \ldots, k+1\}$ and set $D_j(p) := \Psi_j^{-1}(\Psi_j(p))$ for $p \in M$. Then the map $p \mapsto (S(p), D_j(p))$ is a bijection between M and the set $\{(S, D_j) \mid S$ soul of M, D_j fiber of $\Psi_j\}$. Moreover, $D_j(p), D_j(q)$ are equidistant for all $p, q \in M$, and $x \in D_j(p)$, $y \in D_j(q)$ satisfy $|xy| = |D_j(p)D_j(q)|$ if and only if $S(x) = S(y)$.

Proof. All these facts follow from Theorem 3.12 and its proof; compare also the proof of Lemma 3.14. □

Since the sets D_j will be used later, we assign a special name to them.

3.17 Definition. For each $j \in \{1, \ldots, k+1\}$ the set $D_j(p)$ defined as in Corollary 3.16 is called the *(j-th) dual fiber* through $p \in M$.

In the subsequent section we will prove that in fact all dual fibers through any point p coincide. The name will be justified later by the fact that all dual fibers are isometric, providing the second factor of the metric product $M \cong S \times D$ in the Splitting Theorem.

3.6 The dual fibers

In order to prove that j-th dual fibers coincide for different values of j, we examine the canonical projections along souls. They turn out to be locally Lipschitz almost everywhere, which implies that Lipschitz paths are mapped onto Lipschitz paths. Moreover, Rademacher's Theorem can be applied. For that reason, the following result plays a crucial role.

3.18 Proposition. Let $j \in \{1, \ldots, k+1\}$ and D_j some j-th dual fiber. Let $\pi_{D_j} \colon M \to D_j$ denote the projection along souls onto D_j, i.e. $\pi_{D_j}(p) = S(p) \cap D_j$, $p \in M$. Then π_{D_j} is locally a Lipschitz map near almost all points. More precisely, if $p \in M$ is a regular point of the Alexandrov space $S(p)$, there is an open neighborhood $U \mathring{\subseteq} S(p)$ such that $\pi_{D_j}\big|_{\Psi_j^{-1}(\Psi_j(U))}$ is Lipschitz.

Proof. Let $p \in M$ be a regular point of its soul $S(p)$, i.e. p is $(n-k, \delta)$-strained in $S(p)$ for all $\delta > 0$, see Definition 1.11 on page 12. Choose some small $\delta > 0$ and an open neighborhood $V \mathring{\subseteq} S(p)$ of p small enough such that all points $x \in V$ are also $(n-k, \delta)$-strained with the same points $b_1, \ldots, b_{2(n-k)}$ of the strainers. We set $h_i(x) := \frac{1}{2}|b_i x|^2$ for $i = 1, \ldots, 2(n-k)$, $x \in S(p)$. By construction, there is a global constant $C > 0$ such that $|\nabla_x h_i| \geq C$ $\forall x \in V$, $i = 1, \ldots, 2(n-k)$. The functions h_i are 1-concave, and so are all convex combinations $\sum \beta_i h_i$ (where $\sum \beta_i = 1$, $\beta_i \geq 0$). If $x \in V$ is regular and fulfills $|\Uparrow_x^{b_i}| = 1$ $\forall i$, then all differentials $d h_i$ are linear and hence $\nabla_x\big(\sum \beta_i h_i\big) = \sum \beta_i \nabla_x h_i$ for all convex combinations. By Proposition 2.2 on page 26, this holds for almost all points $x \in V$, and since they are $(n-k, \delta)$-strained, the

Chapter 3. The Splitting Theorem

positive constant C from above can be chosen such that $\left|\nabla_x\left(\sum \beta_i h_i\right)\right| \geq C$ holds for almost all $x \in V$ and all convex combinations. Moreover, for all such points x and any direction $\xi \in \Sigma_x S(p)$, there exist $\beta_i \geq 0$, $\sum \beta_i = 1$ such that $\nabla_x\left(\sum \beta_i h_i\right) = \sum \beta_i \nabla_x h_i$ has the given direction ξ.

Now let $U \subseteq V$ be an open neighborhood of p small enough such that $xy \subseteq V$ for all $x, y \in U$. The results from above together with Proposition 2.1 on page 23 imply the following fact: For any two points $x, y \in U$ there is a shortest path γ with endpoints in U arbitrarily close to x and y, respectively, such that almost every tangent vector of γ coincides with the gradient of a suitable convex combination $\sum \beta_i h_i$. To be precise, the equality of the vectors holds only up to length; we reparametrize γ to get full equality. Since all gradient lengths are bounded below by C, the reparametrized path γ is defined on an interval of length at most $T < \infty$. Here, T is a universal constant valid for all such paths γ, independent of the given points $x, y \in U$.

The proof of the Proposition is now carried out in the subsequent three steps.

Step 1. For any $x, y \in U$ the projection along souls from $D_j(x)$ onto $D_j(y)$ is e^T-Lipschitz.

In order to show this, we define the functions $\hat{h}_i \colon M \to \mathbb{R}$, $q \mapsto \frac{1}{2}|q\, D_j(b_i)|^2$ for $i = 1, \ldots, 2(n-k)$. It follows that also the functions $\hat{h}_i$ are 1-concave. Furthermore, according to Corollary 3.16 on the previous page, we have that $\hat{h}_i\big|_{S(p)} = h_i$. Since all souls are isometric, a respective statement holds in each soul. Now we assume for the moment that x and y can be connected by some shortest path γ as above. Thus, almost every tangent vector of γ coincides with the gradient of some convex combination $\sum \beta_i \hat{h}_i$. For any other soul we get the same result, and since the functions $\hat{h}_i$ are defined on the entire space M, Lemma 1.20 on page 16 implies (compare also Lemma 1.23) that the following holds: If $x_1, x_2 \in D_j(x)$ and $y_\ell = S(x_\ell) \cap D_j(y)$, $\ell = 1, 2$, then $|y_1 y_2| \leq e^T |x_1 x_2|$. If x and y cannot be connected directly by such path γ, there are adequate points in each neighborhood of x and y. Therefore, the Lipschitz constant stays the same.

Step 2. For any $x \in U$ the projection along souls from $\Psi_j^{-1}(\Psi_j(U))$ onto $D_j(x)$ is $2e^T$-Lipschitz.

This follows easily from step 1. Indeed, let $q, r \in \Psi_j^{-1}(\Psi_j(U))$ and project r onto $D_j(q)$, i.e. r is mapped onto $r' := S(r) \cap D_j(q)$. We obtain that

$$|qr'| \leq |qr| + |rr'| \leq |qr| + |D_j(r)\, D_j(q)| \leq 2|qr|$$

and hence the statement follows by step 1.

Step 3. The projection along souls from $\Psi_j^{-1}(\Psi_j(U))$ onto D_j is Lipschitz.

For the proof take some shortest path τ from p to D_j, i.e. τ lies in $S(p)$ and ends at $S(p) \cap D_j$. Let $z \in \tau$ be an interior point close enough to p such that $z \in U$. According

3.6. The dual fibers

to step 2, the projection along souls from $\Psi_j^{-1}(\Psi_j(U))$ onto $D_j(z)$ is Lipschitz. Hence, it is sufficient to show that also the projection along souls from $D_j(z)$ onto D_j is Lipschitz. It is clear that the gradient flow of the function $x \mapsto \frac{1}{2}|xp|^2$ pushes z along τ onto $S(p) \cap D_j$ and the same is true via isometric copies in all souls. Like in step 1, the gradient flow of the function $x \mapsto \frac{1}{2}|x\,D_j(p)|^2$ has the identical action on each soul. Moreover, since $z \neq p$, the flow pushes $D_j(z)$ onto D_j in finite time. By Lemma 1.23 on page 17, the gradient flow is Lipschitz and so is the projection along souls. $\square$

3.19 Corollary. *For any j-th dual fibers D_j, D_j' the projection along souls from D_j onto D_j' is a homeomorphism.*

Proof. Let $p \in M$ be regular in its soul $S(p)$. By Proposition 3.18, the projections $D_j(p) \to D_j$ and $D_j(p) \to D_j'$ are in particular continuous, hence (by compactness and bijectivity) homeomorphisms. $\square$

3.20 Remark. The fact that points $p \in M$ which are regular in its soul $S(p)$ form a set of full measure in M follows by Theorem 3.12 on page 53 and Rademacher's Theorem 1.15 on page 15. Indeed, each submetry Ψ_j is differentiable almost everywhere with linear differential. The points where this is true are as requested.

Lytchak's results in [Lyt 02] about submetries imply the following statements.

3.21 Lemma. *For $p \in M$ and $j \in \{1,\ldots,k+1\}$ let $S := S(p)$ and $D := D_j(p)$. Then the tangent cones T_pS and T_pD are convex subsets of T_pM and therefore Alexandrov spaces. Their Hausdorff dimensions are $\dim_H T_pS = n - k$ and $\dim_H T_pD = k$. In addition, we have that $T_pS = \{u \in T_pM \mid \langle u,v\rangle = 0 \quad \forall v \in T_pD\}$.*

Proof. Since $S \subseteq M$ is a convex subset of Hausdorff dimension $n - k$, the same holds for the tangent cone T_pS. According to Theorem 3.12 on page 53, the map $\Psi_i \colon M \to S_i := F_1 \cap \ldots \cap \widehat{F_i} \cap \ldots \cap F_{k+1}$ is a submetry. By [Lyt 02, Proposition 5.1] it induces a homogeneous submetry $d_p\Psi_i \colon T_pM \to T_pS_i$ of the tangent cones. Again Theorem 3.12 implies that $d_p\Psi_i\big|_{T_pS}$ is an isometry. Therefore, if $u \in T_pS$, then $d_p\Psi_i$ preserves its length (this property of u is called *horizontal* by Lytchak). [Lyt 02, Lemma 5.3] and Corollary 3.16 imply that the converse is also true, i.e. $u \in T_pM$ fulfills $|d_p\Psi_i(u)| = |u|$ if and only if $u \in T_pS$. Hence, $d_p\Psi_i$ is a regular submetry as defined in [Lyt 02, Definition 6.4]. Now [Lyt 02, Korollar 7.5] gives that $\dim_H T_pD = k$. The cone $T_pD \subseteq T_pM$ coincides with the preimage $\left(d_p\Psi_i\right)^{-1}(o)$ by [Lyt 02, Proposition 5.2]. The latter is a convex subset by [Lyt 02, Proposition 6.4(1)], while (2) implies that $d_p\Psi_j$ preserves the lengths of precisely those vectors in the cone $K(P)$ over the polar set $P \subseteq \Sigma_p$ of Σ_pD, i.e. $P = \{\xi \in \Sigma_p \mid |\xi\,\Sigma_pD| \geq \frac{\pi}{2}\}$. Therefore we obtain that $T_pS = K(P)$ and hence $T_pS = \{u \in T_pM \mid \langle u,v\rangle \leq 0 \quad \forall v \in T_pD\}$.

Now let $q \in S$, $q \neq p$ and $\xi \in \Sigma_p D$. Assume, by way of contradiction, that $|\xi \uparrow_p^q| > \frac{\pi}{2}$. We choose in D a sequence $p_m \xrightarrow{m \to \infty} p$ such that $\uparrow_p^{p_m} \to \xi$ and set $q_m := S(p_m) \cap D_j(q)$. Corollary 3.19 implies that $q_m \to q$. By passing to a subsequence we may assume that the directions $\uparrow_q^{q_m}$ converge to some $\zeta \in \Sigma_q D_j(q)$. Now the assumption together with $|\zeta \uparrow_q^p| \geq \frac{\pi}{2}$ (by the previous result) implies that there exists some $N \in \mathbb{N}$ such that $|p_N q_N| > |pq|$. This contradicts Corollary 3.16 on page 57.

By lower semi-continuity of angles the result holds for all directions in $\Sigma_p S$. In other words, we obtain that $\langle u, v \rangle = 0 \ \ \forall u \in T_p S \ \forall v \in T_p D$. Thus, the Lemma is proved. $\square$

We are now able to prove that we do not have to distinguish j-th dual fibers, but there are unique dual fibers through all points.

3.22 Theorem. *For any $p \in M$ holds $D_i(p) = D_j(p) =: D(p) \ \ \forall i, j \in \{1, \ldots, k+1\}$.*

Proof. The assertion is proved for $i = 1$, $j = 2$ and almost all points p. Let $p \in M$ such that it is regular in the soul $S(p)$. Let $S \subseteq M$ be a soul and set $q_1 := S \cap D_1(p)$, $q_2 := S \cap D_2(p)$. Assume, by way of contradiction, that $q_1 \neq q_2$. Since all dual fibers are path connected, q_1 can be assumed to lie arbitrarily close to p. Moreover, by [Lyt 02, Theorem 7.2] the induced metric and the induced intrinsic metric on dual fibers are locally bi-Lipschitz equivalent. Thus, there exists a Lipschitz path $\gamma_1 \colon [a, b] \to D_1(p)$ from p to q_1. By choosing q_1 close enough to p, we can apply Proposition 3.18 on page 57 and obtain that the projection along souls onto $D_2(p)$ maps γ_1 onto a Lipschitz path $\gamma_2 \subseteq D_2(p)$ from p to q_2.

According to [PP 95, Proposition 2.1(a)], the left and right tangent vectors $\gamma_\ell^+(t), \gamma_\ell^-(t)$ exist and are opposite for almost all $t \in [a, b]$, $\ell = 1, 2$. Thus, the function $L \colon [a, b] \to \mathbb{R}$, $t \mapsto |\gamma_1(t) \gamma_2(t)|$ is differentiable for almost all t. Moreover, for such t we have that $L'(t) = 0$ by Lemma 3.21. Since L is Lipschitz, in particular absolutely continuous, L is constant. This implies that $0 \neq |q_1 q_2| = L(a) = L(b) = |pp| = 0$, contradiction.

Thus, the Theorem is proved for the dual fibers through almost all points $p \in M$. The equality of i-th and j-th dual fibers carries over to their Hausdorff limits (being also i-th and j-th dual fibers, respectively, by Theorem 3.12) and hence, the statement is proved for all dual fibers. $\square$

3.23 Corollary. *Let $p \in M$. Then the gradient curves of all functions $f_1, \ldots, f_{k+1}$ starting at p lie in $D(p)$. The same holds for all shortest paths from p to the boundary strata $F_1, \ldots, F_{k+1}$. In particular, the corresponding directions are all contained in the $(k-1)$-dimensional Alexandrov space $\Sigma_p D$.*

Proof. This follows immediately from Theorem 3.22 and Lemma 3.21. Recall that a shortest path from p to F_i with base point $q \in F_i$ is contained in the gradient curve α_q of the function f_i. $\square$

3.7 Equidistance of the souls

In the next section we examine the lengths of such shortest paths to the boundary strata more closely.

3.7 Equidistance of the souls

In this section we prove that the souls of M are equidistant, which concludes the proof of the Splitting Theorem. The previous results already imply that certain souls are equidistant, namely along shortest paths to the boundary strata F_i. More precisely, we obtain the following result.

3.24 Proposition. *Let $i \in \{1, \ldots, k+1\}$ and denote by α_x the gradient curve of f_i starting at $x \in M$. Then for each soul $S \subseteq M$ and $t \geq 0$ we have that*

$$L\left(\alpha_p\big|_{[0,t]}\right) = L\left(\alpha_q\big|_{[0,t]}\right) \quad \forall p, q \in S.$$

Proof. The proof is carried out for $i = 1$. Let $p, q \in S$ and let γ be some shortest path from p to q. Choose two distinct interior points $p_0, q_0 \in \gamma$. We claim that $|\nabla_{p_0}(f_1)| = |\nabla_{q_0}(f_1)|$.

If $S \subseteq A_1$, there is nothing to prove. Thus, we may assume that $S \subseteq F_1$, because otherwise we could proceed in the superlevel set $f_1^{-1}([f_1(S), a_1])$. If there is some $j \in \{2, \ldots, k+1\}$ with $S \subseteq F_j$, the direction ξ of the gradient $\nabla_{p_0}(f_1)$ satisfies $\xi \in \Sigma_{p_0} F_j$. This implies that $\nabla_{p_0}(f_1)$ stays the same if we consider the doubling $\bar{M}$ obtained by gluing along $F_2 \cup \ldots \cup F_{k+1}$. In $\bar{M}$ we have that $\bar{\Sigma}_{p_0} \bar{F}_1 = \partial \bar{\Sigma}_{p_0}$. The direction ξ is the (unique) one at maximal distance to $\partial \bar{\Sigma}_{p_0}$. The analog statement holds for q_0 and since p_0, q_0 are interior points of γ, Petrunin's parallel transportation implies that $\bar{\Sigma}_{p_0} \cong \bar{\Sigma}_{q_0}$, see [Pet 98, Theorem 1.1A]. Hence, the claim $|\nabla_{p_0}(f_1)| = |\nabla_{q_0}(f_1)|$ follows.

Now let $t \geq 0$. According to Theorem 3.12 on page 53, the pushed soul $\Phi^t_{f_1}(S)$ is again a soul of M and is isometric to S via the gradient flow. In particular, $\Phi^t_{f_1}(\gamma)$ is a shortest path from $\Phi^t_{f_1}(p)$ to $\Phi^t_{f_1}(q)$ containing the distinct interior points $\Phi^t_{f_1}(p_0), \Phi^t_{f_1}(q_0)$. For these points the claim from above holds, too. This implies that the gradient curves starting at p_0 and q_0, respectively, satisfy $L\left(\alpha_{p_0}\big|_{[0,t]}\right) = L\left(\alpha_{q_0}\big|_{[0,t]}\right)$.

We choose now sequences p_n, q_n of interior points of γ such that $p_n \to p$ and $q_n \to q$. The argument from above carries over to all p_n, q_n, and we obtain sequences of converging gradient curves with equal lengths. According to [Pet 07, Lemma 2.1.5] and its proof, the limits are the gradient curves α_p and α_q, respectively, having the same lengths on each subinterval. □

3.25 Corollary. *For each $i \in \{1, \ldots, k+1\}$ the following holds: Let $p \in M \setminus F_i$ and let γ be some shortest path from p to F_i. Then for any dual fiber D the canonical projection $\gamma \to D$, $\gamma(t) \mapsto S(\gamma(t)) \cap D$ is a shortest path from $q := S(p) \cap D$ to F_i.*

Proof. Let $p_0 \in F_i$ be the endpoint of γ and set $q_0 := S(p_0) \cap D$. We denote by $\gamma^{\leftarrow} \colon [0, f_i(p)] \to M$, $t \mapsto \gamma(f_i(p) - t)$ the curve coinciding with γ, but with reversed parametrization. Then the (unit speed) curve $\gamma^{\leftarrow}$ coincides with the gradient curve $\alpha_{p_0}\big|_{[0, f_i(p)]}$ of the function f_i. According to Proposition 3.24, it has the same length as the gradient curve $\alpha_{q_0}\big|_{[0, f_i(p)]} =: \delta$. Hence, $\delta^{\leftarrow}$ is a shortest path from q to F_i and by Theorem 3.12 on page 53, the curve $\delta^{\leftarrow}$ coincides with the canonical projection of γ along souls. □

In order to show in the end that all souls are equidistant subsets of M, we will use the projection along souls to transport some shortest path between two souls into any dual fiber. The key point is to show that such projections are 1-Lipschitz maps. This in turn will be proved using the differential near souls with special properties.

3.26 Lemma. *Let $p \in M \setminus \partial M$ such that $|\Uparrow_p^{F_i}| = 1$ $\forall i \in \{1, \ldots, k+1\}$. Then $|\Uparrow_q^{F_i}| = 1$ $\forall i \in \{1, \ldots, k+1\}$ holds for all $q \in S(p)$, too. Moreover, we have that $\left|\measuredangle_q^{F_i} \uparrow_q^{F_j}\right| = \left|\measuredangle_p^{F_i} \uparrow_p^{F_j}\right|$ $\forall i, j \in \{1, \ldots, k+1\}$.*

Proof. Choose some $q \in S(p)$. The first statement is an immediate consequence of Corollary 3.25 on the previous page. This Corollary also implies, together with Theorem 3.22 on page 60, that the shortest path from q to any F_i coincides with the projection along souls of the shortest path from p to F_i into the unique dual fiber $D(q)$. Therefore, the directional derivative $d_p f_i(\uparrow_p^{F_j})$ coincides with $d_q f_i(\uparrow_q^{F_j})$ for all $i, j \in \{1, \ldots, k+1\}$. This implies the second statement. □

Before we state and prove the next proposition, we provide the following technical lemma.

3.27 Lemma. *Let $1 \le \ell \le k$ and $v_1, \ldots, v_{\ell+1} \in \mathbb{R}^k$ such that the following holds:*

(i) *$v_1, \ldots, v_{\ell+1}$ are minimal linearly dependent, i.e. the vectors of any subcollection $v_1, \ldots, \hat{v}_i, \ldots, v_{\ell+1}$ are linearly independent for all $i \in \{1, \ldots, \ell+1\}$;*

(ii) *$\measuredangle(v_i, v_j) \ge \frac{\pi}{2}$ $\forall i \ne j$.*

Let $V := \mathrm{span}(v_1, \ldots, v_{\ell+1})$ and $W := \{w \in \mathbb{R}^k \mid \langle v_i, w \rangle \le 0 \ \forall i \in \{1, \ldots, \ell+1\}\}$. Then W is a vector subspace fulfilling $\dim W = k - \ell$ and $W = V^{\perp}$.

Proof. The main part is to prove that $V = \left\{\sum_{i=1}^{\ell+1} \alpha_i v_i \,\middle|\, \alpha_i \ge 0\right\}$. We will use induction over ℓ. The base step $\ell = 1$ is clear, hence let $\ell \ge 2$. In the following we work in $V = \mathbb{R}^{\ell}$ and we assume without loss of generality that $v_{\ell+1} = (-1, 0, \ldots, 0)$. For $i = 1, \ldots, \ell$ we define w_i as the orthogonal projection of v_i onto $v_{\ell+1}^{\perp}$, i.e. w_i is obtained by setting the first component of v_i to 0. According to our assumptions, $w_i \ne 0$ for $i = 1, \ldots, \ell$. We claim that $w_1, \ldots, w_{\ell} \in v_{\ell+1}^{\perp} \cong \mathbb{R}^{\ell-1}$ fulfill the induction hypothesis.

3.7. Equidistance of the souls

It is clear that $w_1, \ldots, w_\ell$ are linearly dependent. Now assume that, say, $w_2, \ldots, w_\ell$ are already linearly dependent. Then there is a non-trivial linear combination $\sum_{i=2}^{\ell} \lambda_i w_i = 0$. By definition of the w_i, this implies that $\sum_{i=2}^{\ell} \lambda_i v_i = \lambda v_{\ell+1}$ for some $\lambda \in \mathbb{R}$. Thus, $v_2, \ldots, v_{\ell+1}$ are linearly dependent, which contradicts the assumptions.

In addition we have that $\angle(w_i, w_j) \geq \frac{\pi}{2}$ $\forall i \neq j$. Indeed, for each $i = 1, \ldots, \ell$ the unit vectors $\frac{v_i}{|v_i|}, \frac{v_{\ell+1}}{|v_{\ell+1}|} \in \mathbb{S}^{\ell-1}$ can be connected by a unique shortest path in $\mathbb{S}^{\ell-1}$. Its intersection point with $v_{\ell+1}^\perp$ coincides with the unit vector $\frac{w_i}{|w_i|}$. Now it is easy to see that $\angle(w_i, w_j) < \frac{\pi}{2}$ would imply that $\angle(v_i, v_j) < \frac{\pi}{2}$.

Hence, by induction assumption, we have that $v_{\ell+1}^\perp = \left\{ \sum_{i=1}^{\ell} \alpha_i w_i \,\middle|\, \alpha_i \geq 0 \right\}$. It follows immediately that the half space $\{(x_1, \ldots, x_\ell) \in \mathbb{R}^\ell \,|\, x_1 \leq 0\}$ is contained in the convex set $\left\{ \sum_{i=1}^{\ell+1} \alpha_i v_i \,\middle|\, \alpha_i \geq 0 \right\}$. Thus, the entire space V is contained if and only if there is some $v_i \in \{(x_1, \ldots, x_\ell) \in \mathbb{R}^\ell \,|\, x_1 > 0\}$. This is equivalent to $w_i \neq v_i$. But the latter is satisfied for at least one index i, since otherwise $v_1, \ldots, v_\ell$ would be linearly dependent.

Now we prove the assertion of the Lemma. Let $w \in W$ and $j \in \{1, \ldots, \ell+1\}$. By the previous result, there are coefficients $\alpha_1, \ldots, \alpha_{\ell+1} \geq 0$ such that $-v_j = \sum_{i=1}^{\ell+1} \alpha_i v_i$. It follows that

$$-\langle v_j, w \rangle = \left\langle \sum_{i=1}^{\ell+1} \alpha_i v_i, w \right\rangle = \sum_{i=1}^{\ell+1} \alpha_i \langle v_i, w \rangle \leq 0$$

and therefore $\langle v_j, w \rangle = 0$. We obtain that $W = \left\{ w \in \mathbb{R}^k \,\middle|\, \langle v_i, w \rangle = 0 \ \forall i \in \{1, \ldots, \ell+1\} \right\}$ which yields the desired results. $\square$

We are now able to examine the differential of the projection along souls almost everywhere.

3.28 Proposition. *Let D be some dual fiber and $\pi_D \colon M \to D$, $x \mapsto S(x) \cap D$ the projection along souls onto D. Then for almost all $p \in M$ the differential $d_p \pi_D$ exists and is 1-Lipschitz.*

Proof. According to Proposition 3.18 on page 57, the projection π_D is locally Lipschitz near almost all points. Hence, it is differentiable almost everywhere with linear differential, compare Theorem 1.15 on page 15. More precisely, we consider π_D as a map into M (thus, the target space is Alexandrov) with image in D, where D is equipped with the induced metric. Then for almost all $p \in M$ we have the following.

- $T_p M$ is isometric to $\mathbb{R}^n$.

- $T_{\pi_D(p)} M$ is isometric to $\mathbb{R}^m \times C$, where $C \in \text{Alex}^{n-m}(0)$ is a cone.

- $d_p \pi_D$ is a linear map with image in the $\mathbb{R}^m$-factor of $T_{\pi_D(p)} M$.

Moreover, the Euclidean cone T_pM can be assumed to split into the factors $T_pS(p) \cong \mathbb{R}^{n-k}$ and $T_pD(p) \cong \mathbb{R}^k$. Indeed, since to each vector $v \in T_pM$ there exists the opposite vector $-v \in T_pM$, it follows from Lemma 3.21 on page 59 that $T_pS(p) \cong \mathbb{R}^{n-k}$. According to the proof of that Lemma, $T_pD(p) = (d_p\Psi_i)^{-1}(0)$ for each $i \in \{1,\ldots,k+1\}$. All maps Ψ_i are 1-Lipschitz, hence $d_p\Psi_i$ is linear at almost all $p \in M$. If we take such a point p, it follows that $T_pD(p) \cong \mathbb{R}^k$.

Finally, we may assume that each shortest path from p to any F_i can be extended beyond p, since this is true for almost all points according to Proposition 2.2 on page 26. For $i = 1,\ldots,k+1$ we set $b_i := \uparrow_p^{F_i} \in T_pD(p)$. According to Lemma 2.12 on page 33 we have that

$$|b_ib_j| \geq \frac{\pi}{2} \quad \forall i \neq j.$$

In addition, Corollary 3.25 on page 61 and Lemma 3.26 imply that

$$\langle b_i, b_j \rangle = \langle d_p\pi_D(b_i), d_p\pi_D(b_j) \rangle \quad \forall i,j \in \{1,\ldots,k+1\}.$$

Now let $v \in T_pM$. By the orthogonal splitting $T_pM = T_pS(p) \times T_pD(p)$ it is sufficient to consider $v \in T_pD(p)$. If there are coefficients $\beta_i \in \mathbb{R}$ such that $v = \sum_{i=1}^{k+1}\beta_ib_i$, linearity of $d_p\pi_D$ implies the following:

$$\begin{aligned}|v|^2 &= \langle v,v \rangle = \sum_{i,j}\beta_i\beta_j\langle b_i, b_j\rangle = \sum_{i,j}\beta_i\beta_j\langle d_p\pi_D(b_i), d_p\pi_D(b_j)\rangle \\ &= \langle d_p\pi_D(v), d_p\pi_D(v)\rangle = |d_p\pi_D(v)|^2\end{aligned}$$

Hence, the Proposition is proved in the case of $\mathrm{span}(b_1,\ldots,b_{k+1}) = T_pD(p)$.

In the general case we choose some minimal linearly dependent subcollection of $\{b_i\}$ in the sense of Lemma 3.27. Assume by renumbering, that $b_1,\ldots,b_{\ell+1}$ is such subcollection, where $1 \leq \ell \leq k$. Let V and W be given as in Lemma 3.27, hence $T_pD(p)$ splits orthogonally as $T_pD(p) = V \oplus W$. According to the results from above, it is sufficient to consider $v \in W$.

For $i = 1,\ldots,\ell$ we shift the boundary strata F_i to p, i.e. we consider the intersection $M' := f_1^{-1}([f_1(p), a_1]) \cap \ldots \cap f_\ell^{-1}([f_\ell(p), a_\ell])$ of superlevel sets. Each boundary stratum F_i is perpendicular to the corresponding shortest path from p to F_i and the same holds in superlevel sets. It also holds in M', since all these shortest paths are extendable beyond p. Thus, the tangent cone T_pM' coincides with the set $\{v \in T_pM \mid \langle v, b_i \rangle \leq 0 \; \forall i \in \{1,\ldots,\ell\}\}$. No collapse occurs, because $b_1,\ldots,b_\ell$ are linearly independent and therefore T_pM' has full dimension.

3.7. Equidistance of the souls

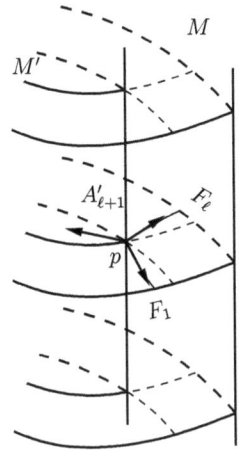 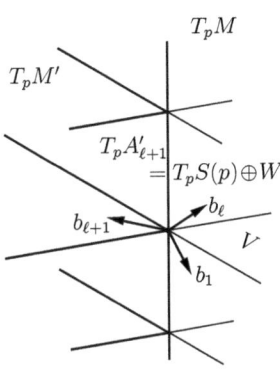

We claim that $f_{\ell+1}\big|_{M'}$ attains its maximum at $p \in M'$. Indeed, let $w \in T_pM'$ such that $\langle w, b_{\ell+1}\rangle \leq 0$. Then $w \in T_pS(p) \oplus W$, which implies that in fact $\langle w, b_{\ell+1}\rangle = 0$. The claim follows by concavity of $f_{\ell+1}$. By the claim, the vector $v \in W \subseteq T_pM'$ can be considered as $v \in W \subseteq T_pA'_{\ell+1}$, where $A'_{\ell+1} = \max_{x \in M'} f_{\ell+1}(x)$. More precisely, we have that $T_pA'_{\ell+1} = T_pS(p) \oplus W$. By induction assumption, the statement of the Proposition is proved for the set $A'_{\ell+1}$ and hence, $|d_p\pi_D(v)| = |v|$. This completes the proof. □

3.29 Theorem. *For each dual fiber D, equipped with the induced metric, the projection along souls $\pi_D \colon M \to D$, $x \mapsto S(x) \cap D$ is 1-Lipschitz.*

Proof. Choose an arbitrary dual fiber D. Let $p, q \in M$ be distinct points and $\varepsilon > 0$. Choose $\delta > 0$ small enough such that $\pi_D(B_\delta(x)) \subseteq B_\varepsilon(\pi_D(x))$ for $x = p$ and $x = q$ (recall Corollary 3.19 on page 59). According to Proposition 3.28 and Proposition 2.1 on page 23, there are points $\hat{p} \in B_\delta(p)$, $\hat{q} \in B_\delta(q)$ such that for almost all points $x \in \hat{p}\hat{q}$ the differential $d_x\pi_D$ is 1-Lipschitz. This implies that $|\pi_D(\hat{p})\,\pi_D(\hat{q})| \leq |\hat{p}\hat{q}|$. Thus, we obtain the following:

$$\begin{aligned}|\pi_D(p)\,\pi_D(q)| &\leq |\pi_D(p)\,\pi_D(\hat{p})| + |\pi_D(\hat{p})\,\pi_D(\hat{q})| + |\pi_D(\hat{q})\,\pi_D(q)| \\ &\leq \varepsilon + |\hat{p}\hat{q}| + \varepsilon \\ &\leq 2\varepsilon + 2\delta + |pq|\end{aligned}$$

By choosing ε and hence δ arbitrarily small, the result follows. □

3.30 Corollary. *The souls form an equidistant fibration of M. All dual fibers are isometric and form convex subsets of M.*

Proof. Let S_1, S_2 be souls and $p \in S_1$, $q \in S_2$ such that $|S_1 S_2| = |pq|$. By Theorem 3.29, the shortest path pq can be transported via projection along souls into any dual fiber D and satisfies $L(\pi_D(pq)) \leq |pq|$. Thus, equality follows, which proves the first statement and therewith also the second statement. □

Now the proof of the Splitting Theorem can be completed.

Proof of the Splitting Theorem 3.1 on page 43. We pick arbitrarily some soul S and some dual fiber D. Since the set of souls is equidistant and so is the set of dual fibers, the proof that M is isometric to the product $S \times D$ (equipped with the product metric) works exactly as in the two strata case, see the proof of Theorem 3.3 on page 45. Indeed, in M the Pythagorean Theorem holds up to an error of order $o(h)$ if h is some side length of a rectangular triangle. Since souls and dual fibers intersect perpendicularly, we obtain the canonical isometry $\pi_S \times \pi_D \colon M \to S \times D$, where π_S, π_D are the canonical projections. □

CHAPTER 4

Consequences of the Splitting Theorem

In this chapter we collect some easy consequences of the Splitting Theorem. As before, we may consider the factors D and S as subsets of M, i.e. we may identify D and S with $\{x\} \times D$ and $S \times \{y\}$, respectively, for arbitrarily chosen $x \in S$, $y \in D$.

4.1 Corollary. *Under the assumptions of the Splitting Theorem, the sets $F_i \cap D$, $i = 1, \ldots, k+1$ form a stratification of ∂D. Moreover, if $F \subseteq M$ is a primitive extremal subset of codimension 1 in M, then $F \cap D$ is a primitive extremal subset of codimension 1 in D.*

Proof. Let F be some boundary stratum or one of its primitive components of full dimension. For $p \in M \setminus F$, the endpoint $q \in F$ of any shortest path from p to F lies in the same dual fiber $D(q) = D(p)$. It follows that $F \cap D$ is extremal in D, and by the dimension, it is a boundary stratum. Vice versa, each boundary stratum (or its primitive components) $G \subseteq \partial D$ gives rise to a boundary stratum of M via the product structure, i.e. $G \times S \subseteq \partial M$. This proves all statements of the Corollary. □

4.2 Corollary. *Let everything be as in the Splitting Theorem. Assume, in addition, that $P_1, \ldots, P_\ell \subseteq \partial M$ are the primitive boundary components, i.e. each boundary stratum F_i may consist of several primitive components P_j. (Hence, $\ell \geq k+1$.) Then each nonempty intersection of k primitive components P_j is a soul of M.*

Proof. This follows from the Splitting Theorem and Corollary 4.1. Indeed, if we assume that $P := P_1 \cap \ldots \cap P_k \neq \emptyset$, it induces the intersection $(P_1 \cap D) \cap \ldots \cap (P_k \cap D) = P \cap D$ in the D-factor of M. By Theorem 2.18 on page 37, this intersection is just a point p. Hence, by the product structure, $P = S(p)$ is a soul. □

The Splitting Theorem was formulated for the case that $F_1, \ldots, F_{k+1}$ is a stratification of the boundary. If it is only a subcollection of some stratification, the Splitting Theorem can be applied to the doubling $\bar{M}$ which is obtained by gluing along the inappropriate strata. The question arises, what this construction implies for the original space M. Obviously, the product structure carries over if the strata which are glued away come from boundary strata of S. This is indeed the case and formulated in the following extension of the Splitting Theorem.

4.3 Theorem (Extended Splitting Theorem). *Let $M \in \mathrm{ALEX}^n(0)$ be compact and let $F_1, \ldots, F_{\ell+1}$ be a stratification of ∂M. Assume that there is $1 \leq k \leq \ell$ such that the following holds:*

- $F_1 \cap \ldots \cap F_{k+1} = \emptyset$
- $F_1 \cap \ldots \cap \widehat{F_i} \cap \ldots \cap F_{k+1} \neq \emptyset \quad \forall i \in \{1, \ldots, k+1\}$

Then $M \cong S \times D$, where $S \in \mathrm{ALEX}^{n-k}(0)$ is isometric to $F_1 \cap \ldots \cap \widehat{F_i} \cap \ldots \cap F_{k+1}$ for all $i \in \{1, \ldots, k+1\}$ and $D \in \mathrm{ALEX}^k(0)$. Moreover, for $i = k+2, \ldots, \ell+1$ and all $p \in M$ the intersections $F_i \cap S(p)$ are nonempty and form a stratification of $\partial S(p) \cong \partial S$.

Proof. As mentioned above, if we consider the doubling $\bar{M}$ by gluing along $F_{k+2} \cup \ldots \cup F_{\ell+1}$, we can apply the original Splitting Theorem. Thus, only the last statement has to be proved. Again, we may simplify the setting by passing to $\bar{M}$ and gluing along $F_{k+2} \cup \ldots \cup F_\ell$. Or in other words, we may assume that ∂M consists of the strata $F_1, \ldots, F_{k+1}, F_{\ell+1}$.

Let $p \in F_{\ell+1}$ and $S := S(p)$ be the soul through p. We claim that $F_{\ell+1} \cap S = \partial S$. If S coincides with some intersection $F_1 \cap \ldots \cap \widehat{F_j} \cap \ldots \cap F_{k+1}$ or this is true in some superlevel set $M' \subseteq M$, then $F_{\ell+1} \cap S$ is an extremal subset of M or M', respectively. Hence, it is also extremal in S and it has codimension 1 in S by Theorem 2.18 on page 37. Conversely, according to Proposition 2.21 on page 40, each boundary point $q \in \partial S$ has to lie in some boundary stratum distinct from $F_1, \ldots, F_{k+1}$. Hence, $q \in F_{\ell+1}$, and the claim is proved in the current case.

If S cannot be written as an intersection of boundary strata, there is some index i such that $S \subseteq A_i$ (inside M or inside some superlevel set $M' \subseteq M$; the set A_i is defined as in Chapter 3). Like in Theorem 3.4 on page 47, it turns out that A_i satisfies the assumptions of the Extended Splitting Theorem. Indeed, $F_{\ell+1} \cap A_i$ is the additional boundary stratum of A_i. We therefore can assume by induction on k that the statement is proved inside A_i; the base step is clear, because for $k = 1$ each soul S can be written as a boundary stratum. Thus, we obtain that $F_{\ell+1} \cap S = \partial S$, and the claim is proved.

Since all souls are isometric, all souls have boundary. Again by Proposition 2.21 or by induction assumption for some A_i, this boundary is induced from $F_{\ell+1}$, in other words $\partial S(p) = F_{\ell+1} \cap S(p) \quad \forall p \in M$. □

As an immediate consequence we get a result on the intersecting behaviour of boundary strata.

4.4 Corollary. *Under the assumptions of the Extended Splitting Theorem, we have that*

$$F_1 \cap \ldots \cap \widehat{F_i} \cap \ldots \cap F_{k+1} \cap F_j \neq \emptyset \quad \forall\, i \in \{1, \ldots, k+1\},\, j \in \{k+2, \ldots, \ell+1\}.$$

In particular, if there are two boundary strata which do not intersect, they are primitive.

Proof. The first statement is clear. The second follows from the case $k = 1$. □

The Extended Splitting Theorem enables its iterated use. Indeed, it may turn out that ∂S possesses some stratification which satisfies again the assumptions of the Theorem. The proof of the following result is an example of how this works.

4.5 Theorem. *Let $M \in \text{Alex}^n(0)$ be compact and $\partial M \neq \emptyset$. Then the stratification of ∂M consists of at most $2n$ strata. In addition, if the number of boundary strata equals $2n$, then M is isometric to a product of n intervals, i.e. $M \cong I_1 \times \cdots \times I_n$. In other words, M possesses $2n$ boundary strata if and only if M is a Euclidean cuboid.*

Proof. Let $F_1, \ldots, F_\ell$ be some stratification of ∂M. If $p \in M$ is regular, then Lemma 2.12 on page 33 immediately implies that $\ell \leq 2n$. Now we assume equality. A first consequence is, that each point $p \in M \setminus \partial M$ is regular with $|\Uparrow_p^{F_i}| = 1 \quad \forall\, i \in \{1, \ldots, 2n\}$. Indeed, since for each $p \in M$ there is a noncontracting map $\Sigma_p \to \mathbb{S}^{n-1}$ (compare the paragraph after Definition 1.8 on page 11), this follows again by Lemma 2.12.

Now choose some $p \in M \setminus \partial M$. It follows that there is exactly one index i such that $|\Uparrow_p^{F_1} \Uparrow_p^{F_i}| = \pi$. Assume, without loss of generality, that $i = 2$. We claim that $F_1 \cap F_2 = \emptyset$. For $i = 2, \ldots, 2n$ we define the sets $N_i := \{x \in M \setminus \partial M\, |\, |\Uparrow_x^{F_1} \Uparrow_x^{F_i}| = \pi\}$. The lower semi-continuity of angles implies that all sets N_i are open in $M \setminus \partial M$. More precisely, fix some index i and let $x_m \to x$ be a convergent sequence in $M \setminus \partial M$ satisfying $x_m \notin N_i \quad \forall\, m \in \mathbb{N}$. This induces sequences $y_m \in F_1$, $z_m \in F_i$ such that $|x_m y_m| = |x_m F_1|$ and $|x_m z_m| = |x_m F_i|$. Since all F_j are compact, we may assume that $y_m \to y$ and $z_m \to z$ converge. By definition, we have that $\angle y_m x_m z_m = \frac{\pi}{2} \quad \forall\, m \in \mathbb{N}$, hence semi-continuity of angles implies that $\angle yxz = \frac{\pi}{2}$ or, in other words, $x \notin N_i$. Therefore, $(M \setminus \partial M) \setminus N_i$ is closed and N_i is open in $M \setminus \partial M$.

This implies that $\bigcup_{i \in \{2, \ldots, 2n\}} N_i$ is a disjoint open cover of the convex set $M \setminus \partial M$. Since $p \in N_2$, we obtain that $N_2 = M \setminus \partial M$ and $N_i = \emptyset$ for $i = 3, \ldots, 2n$. It follows that the function $h \colon M \setminus \partial M \to \mathbb{R},\, x \mapsto |xF_1| + |xF_2|$ has directional derivative 0 at each point in each direction. Hence, $h \equiv d := |F_1 F_2|$ is constant and the claim $F_1 \cap F_2 = \emptyset$ is proved.

We apply the Extended Splitting Theorem for $k=1$ and obtain that $M \cong I_1 \times F_1$, where $I_1 = [0,d]$. Moreover, $F_1 \in \text{ALEX}^{n-1}(0)$ has $2n-2$ boundary strata, namely $F_1 \cap F_3, \ldots, F_1 \cap F_{2n}$. By iterating the procedure from above, we get the final result $M \cong I_1 \times \cdots \times I_n$. □

Bibliography

[AB 05] S.B. ALEXANDER, R.L. BISHOP, *A cone splitting theorem for Alexandrov spaces*, Pacific J. Math. **218** (2005), no. 1, 1-16

[BBI 01] D. BURAGO, Y. BURAGO, S. IVANOV, *A Course in Metric Geometry*, Graduate Studies in Mathematics, Vol. 33, American Mathematical Society, Providence, Rhode Island, 2001

[BGP 92] Y. BURAGO, M. GROMOV, G. PERELMAN, *A.D. Alexandrov spaces with curvatures bounded below*, Russian Math. Surveys **47** (1992), 1-58

[CC 96] J. CHEEGER, T. COLDING, *Lower bounds on Ricci curvature and the almost rigidity of warped products*, Ann. of Math. **144** (1996), 189-237

[CE 75] J. CHEEGER, D. EBIN, *Comparison Theorems in Riemannian Geometry*, North Holland, New York, 1975

[CG 71] J. CHEEGER, D. GROMOLL, *The splitting theorem for manifolds of nonnegative Ricci curvature*, J. Diff. Geom. **61** (1971), 119-128

[CG 72] J. CHEEGER, D. GROMOLL, *On the structure of complete manifolds of nonnegative curvature*, Ann. of Math. **96** (1972), 413-443

[GKM 68] D. GROMOLL, W. KLINGENBERG, W. MEYER, *Riemannsche Geometrie im Großen*, Lecture Notes 55, Springer, 1968

[GT 03] D. GROMOLL, K. TAPP, *Nonnegatively curved metrics on $\mathbb{S}^2 \times \mathbb{R}^2$*, Geom. Dedicata **99** (2003), no. 1, 127-136

Bibliography

[GW 71] D. GROMOLL, J.A. WOLF, *Some relations between the metric structure and the algebraic structure of the fundamental group in manifolds of nonpositive curvature*, Bull. Amer. Math. Soc. **77** (1971), no. 4, 545-552

[Gro 81] M. GROMOV, *Curvature, diameter and Betti numbers*, Comment. Math. Helv. **56** (1981), no. 1, 179-195

[GP 93] K. GROVE, P. PETERSEN, *A radius sphere theorem*, Invent. Math. **112** (1993), 577-583

[Gui 00] L. GUIJARRO, *On the metric structure of open manifolds with nonnegative curvature*, Pacific J. Math. **196** (2000), no. 2, 429-444

[Kap 07] V. KAPOVITCH, *Perelman's stability theorem*, Surveys in Differential Geometry, Volume XI: Metric and Comparison Geometry, International Press, 2007

[LS 97] U. LANG, V. SCHRÖDER, *Kirszbraun's theorem and metric spaces of bounded curvature*, Geom. Funct. Analysis **7** (1997), no. 3, 535-560

[LY 72] H.B. LAWSON, S.T. YAU, *Compact manifolds of nonpositive curvature*, J. Diff. Geom. **7** (1972), 211-228

[Lyt 02] A. LYTCHAK, *Submetrien von Alexandrov-Räumen*, Dissertation, published as *Allgemeine Theorie der Submetrien und verwandte mathematische Probleme*, Bonner Math. Schriften **347** (2002)

[Lyt 05] A. LYTCHAK, *Differentiation in metric spaces*, St. Petersburg Math. Journ. **16** (2005), no. 6, 1017-1041

[Lyt 06] A. LYTCHAK, *Open map theorem for metric spaces*, St. Petersburg Math. Journ. **17** (2006), no. 3, 477-491

[Mas 02] Y. MASHIKO, *A splitting theorem for Alexandrov spaces*, Pacific J. Math. **204** (2002), no. 2, 445-448

[Mil 67] A.D. MILKA, *Metric structure of one class of spaces containing straight lines* (in Russian), Ukrain. Geom. Sbornik **4** (1967), 43-48

[Mit 10] A. MITSUISHI, *A splitting theorem for infinite dimensional Alexandrov spaces with nonnegative curvature and its applications*, Geom. Dedicata **144** (2010), no. 1, 101-114

[Ots 97] Y. OTSU, *Differential geometric aspects of Alexandrov spaces*, in Comparison Geometry, Math. Sci. Res. Inst. Publ 30 Cambridge Univ. Press, Cambridge (1997), 135-148

Bibliography

[OS 94] Y. OTSU, T. SHIOYA, *The Riemannian structure of Alexandrov spaces*, J. Diff. Geom. **39** (1994), no. 3, 629-658

[Per 91] G. PERELMAN, *Alexandrov's spaces with curvatures bounded from below II*, Preprint (1991), http://www.math.psu.edu/petrunin/papers/papers.html

[Per 94a] G. PERELMAN, *Proof of the soul conjecture of Cheeger and Gromoll*, J. Diff. Geom. **40** (1994), 209-212

[Per 94b] G. PERELMAN, *Elements of Morse theory on Alexandrov spaces*, St. Petersburg Math. Journ. **5** (1994), no. 1, 205-214

[PP 94] G. PERELMAN, A. PETRUNIN, *Extremal subsets in Alexandrov spaces and a generalised Liberman theorem*, St. Petersburg Math. Journ. **5** (1994), no. 1, 215-227

[PP 95] G. PERELMAN, A. PETRUNIN, *Quasigeodesics and gradient curves in Alexandrov spaces*, Preprint (1995), http://www.math.psu.edu/petrunin/papers/papers.html

[Pet 97] A. PETRUNIN, *Applications of quasigeodesics and gradient curves*, in Comparison Geometry, Math. Sci. Res. Inst. Publ 30 Cambridge Univ. Press, Cambridge (1997), 203-219

[Pet 98] A. PETRUNIN, *Parallel transportation for Alexandrov spaces with curvature bounded below*, Geom. Funct. Analysis **8** (1998), no. 1, 123-148

[Pet 07] A. PETRUNIN, *Semiconcave functions in Alexandrov's geometry*, Surveys in Differential Geometry, Volume XI: Metric and Comparison Geometry, International Press, 2007

[Pla 91] C. PLAUT, *Almost Riemannian spaces*, J. Diff. Geom. **35** (1991), no. 2, 515-537

[Pla 02] C. PLAUT, *Metric spaces of curvature $\leq k$*, Handbook of geometric topology, 819-898, North-Holland, Amsterdam, 2002

[Sha 79] V.A. SHARAFUTDINOV, *Convex sets in a manifold of nonnegative curvature* (in Russian), Mat. Zametki **26** (1979), no. 1, 129-136; English translation in Math. Notes **26** (1979), no. 1, 556-560

[Sie 72] L.C. SIEBENMANN, *Deformation of homeomorphisms on stratified sets*, Comment. Math. Helv. **47** (1972), 123-163

[Str 88] M. STRAKE, *A splitting theorem for open nonnegatively curved manifolds*, Manuscripta Math. **61** (1988), no. 3, 315-325

[Top 64] V.A. TOPONOGOV, *Riemannian spaces which contain straight lines*, Amer. Math. Soc. Transl., Ser. 2, **37** (1964), 287-290

[Wal 88] G. WALSCHAP, *A splitting theorem for 4-dimensional manifolds of nonnegative curvature*, Proc. Amer. Math. Soc. **104** (1988), no. 1, 265-268

[Wil 06] B. WILKING, *Positively curved manifolds with symmetry*, Ann. of Math. **163** (2006), 607-668

[Wil 07a] B. WILKING, *Nonnegatively and positively curved manifolds*, Surveys in Differential Geometry, Volume XI: Metric and Comparison Geometry, International Press, 2007

[Wil 07b] B. WILKING, *A duality theorem for Riemannian foliations in nonnegative sectional curvature*, Geom. Funct. Analysis **17** (2007), no. 4, 1297-1320

[Yim 90] J.-W. YIM, *Space of souls in a complete open manifold of nonnegative curvature*, J. Diff. Geom. **32** (1990), no. 2, 429-455

List of symbols

General

$\mathbb{N}$ / $\mathbb{Z}$	The natural numbers / integers
$\mathbb{Z}_2$	The field of two elements
$\mathbb{R}$ ($\mathbb{R}_+$)	The (nonnegative) real numbers
$\mathbb{R}^n$	The n-dimensional Euclidean space
$\mathbb{S}^n$	The n-dimensional unit sphere
$X_1, \ldots, \widehat{X_i}, \ldots, X_n$	The collection $X_1, \ldots, X_{i-1}, X_{i+1}, \ldots, X_n$, i.e. the symbol marked by $\frown$ is omitted

Metric spaces

$\overline{A}$	The closure of A
$\overset{\circ}{\subseteq}$	open subset
$\approx$	homeomorphic
$\cong$	isometric, i.e. lengths preserving homeomorphic
$A \times B$	The metric product of A and B, endowed with the Euclidean product metric
dim	The topological dimension; equals $\dim_H$ for Alexandrov spaces
$\dim_H$	The Hausdorff dimension

List of symbols

μ_n	The n-dimensional Hausdorff measure
d	The metric of the space
d_p / d_A	The distance to a point p / to a subset A, i.e. $d_p = d(p,\cdot)$ / $d_A = \mathrm{dist}(A,\cdot) = \inf_{x\in A} d(x,\cdot)$
$\|pq\|$	The distance $d(p,q)$ between two points; mostly used if d is strictly intrinsic and hence, $\|pq\|$ is the length of some shortest path pq
$\|AB\|$	The distance $\mathrm{dist}(A,B) = \inf_{x\in A, y\in B} d(x,y)$ between two subsets; mostly used if d is strictly intrinsic and $\|AB\|$ is assumed by the length of some shortest path connecting A and B
d_H (d_{GH})	The (Gromov-) Hausdorff distance
$\mathrm{diam}\, A$	The diameter of A, i.e. $\mathrm{diam}\, A = \sup_{x,y\in A} \|xy\|$
$B_r(p)$ / $\bar{B}_r(p)$	The open / closed ball of radius r centered at p, i.e. the set of points x satisfying $\|xp\| < r$ / $\|xp\| \le r$
$\mathrm{Bd}_B A$	The topological boundary of $A \subseteq B$, see Definition 2.23 on page 41
$K(\Sigma)$ / $\bar{K}(\Sigma)$	The open / closed metric cone over Σ

Comparison geometry

S^n_κ	The n-dimensional simply connected space form of curvature κ
pq	A shortest path from p to q
$\triangle abc$	A triangle made up of fixed shortest paths ab, ac and bc
$\tilde{\triangle}_\kappa abc$	The comparison triangle $\triangle \bar{a}\bar{b}\bar{c} \subseteq S^2_\kappa$ for $\triangle abc$, see Definition 1.1 on page 8
$\angle abc$	The angle between fixed shortest paths ab and bc
$\tilde{\angle}_\kappa abc$	The comparison angle $\angle \bar{a}\bar{b}\bar{c}$, see Definition 1.1 on page 8

Alexandrov spaces

$\mathrm{ALEX}^n(\kappa)$	The class of n-dimensional complete Alexandrov spaces with curvature bounded below by κ, see Definition 1.4 on page 9
Σ_p	The space of directions at p, see Section 1.3 on page 10
$\uparrow^q_p$ / $\Uparrow^q_p$ / $\Uparrow^A_p$	The direction of a fixed shortest path from p to q / the set of all directions from p to q / from p to a subset A

List of symbols

$T_p M$	The tangent cone at p, see Section 1.3 on page 10
o	The apex of a tangent cone, see Definition 1.7 on page 11
$\|v\|$	The length of a tangent vector, see Definition 1.7 on page 11
$\langle v, w \rangle$	The scalar product of two tangent vectors, see Definition 1.7 on page 11
$\log_p$	A logarithm map at p, see Definition 1.8 on page 11
∂M	The boundary of M, see Definition 1.9 on page 12
$d_p f$	The differential of f at p, see Section 1.5 on page 14
$\gamma^+(t) \,/\, \gamma^-(t)$	The right / left tangent vector of a curve γ at $\gamma(t)$, see Definition 1.16 on page 15
$L(\gamma)$	The length of a Lipschitz curve γ, see Definition 1.16 on page 15
$\nabla_p f$	The gradient of f at p, see Definition 1.19 on page 16
α_p	The gradient curve of some function starting at p, see Definition 1.21 on page 17
Φ_f^t	The gradient flow (also called push) of f at time t, see Definition 1.22 on page 17
S_M	The set of singular points in M, see Definition 1.10 on page 12
$\bar{M}$	The doubling of M, see 2.4 on page 28

Chapter 3

That chapter is in a way one long proof and therefore, main definitions and notation are not repeated for each single lemma, proposition or theorem contained. Here we list the relevant symbols.

M	A compact Alexandrov space in $\text{ALEX}^n(0)$
k	The number of intersecting boundary strata of M, where $k \geq 1$
$F_1, \ldots, F_{k+1}$	Boundary strata; more precisely, the elements of a fixed stratification of ∂M
f_i	The distance function d_{F_i}, $i = 1, \ldots, k+1$
a_i	The maximal value of f_i, i.e. $a_i = \max_{p \in M} f_i(p)$, $i = 1, \ldots, k+1$
A_i	The set where f_i attains its maximum, i.e. $A_i = \{p \in M \mid f_i(p) = a_i\}$, $i = 1, \ldots, k+1$
M'	A non-collapsed convex subset of M obtained as (the intersection of) superlevel sets, see Corollary 2.15 on page 36

List of symbols

$F'_1, \ldots, F'_{k+1}$	The boundary strata of some M', see Corollary 2.15 on page 36
$S(p)$	The soul through $p \in M$, see Proposition 3.6 on page 49
Ψ_j	The submetry $\Psi_j \colon M \to F_1 \cap \ldots \cap \widehat{F_j} \cap \ldots \cap F_{k+1}$ according to Theorem 3.12 on page 53, $j = 1, \ldots, k+1$
$D(p)$ $(D_j(p))$	The (j-th) dual fiber through $p \in M$, see Definition 3.17 on page 57

Index

admissible
    function, 20
    map, 20
Alexandrov space, 9

boundary, 12
    stratum, 28
    topological, 41

Compactness Theorem, Gromov's, 13
comparison
    angle, 8
    triangle, 8
Comparison Theorem
    classical, 7
    extended, 10
    Toponogov's, 9
concave
    λ-, 15
    semi-, 16
curvature bound (lower), 8

differential, 14
direction, 10
directional derivative, 14

distance function, 14
doubling, 28
Doubling Theorem, 28
dual fiber, 57

extremal set, 18
    primitive, 19

gradient, 16
    curve, 17
    flow / push, 17

length
    of a Lipschitz curve, 15
    of a vector, 11
Local Fibration Theorem, 21

MCS-space, 19
$\widetilde{\text{MCS}}$-space, 21
Morse Lemma, *see* Local Fibration Theorem

Rademacher's Theorem, 14
regular point, 12
    for admissible function, 20

semi-continuity
    of angles, 8
    of the gradient, 16

Sharafutdinov retraction, 3
singular point, 12
soul, 49
Soul Theorem, 3, 33
space of directions, 10
Splitting Theorem, 43
    Extended, 68
Stability Theorem, 22
strained, 12
stratification of ∂M, 28
submetry, 53

tangent
    cone, 10
    vector, 11
        right / left, 15

vector, *see* tangent vector

I want morebooks!

Buy your books fast and straightforward online - at one of world's fastest growing online book stores! Environmentally sound due to Print-on-Demand technologies.

Buy your books online at
www.morebooks.shop

Kaufen Sie Ihre Bücher schnell und unkompliziert online – auf einer der am schnellsten wachsenden Buchhandelsplattformen weltweit! Dank Print-On-Demand umwelt- und ressourcenschonend produziert.

Bücher schneller online kaufen
www.morebooks.shop

KS OmniScriptum Publishing
Brivibas gatve 197
LV-1039 Riga, Latvia
Telefax: +371 686 204 55

info@omniscriptum.com
www.omniscriptum.com

Printed by Books on Demand GmbH, Norderstedt / Germany